THE COUNTINGBURY TALES
FUN WITH MATHEMATICS

THE COUNTINGBURY TALES
FUN WITH MATHEMATICS

Miguel de Guzmán
Universidad Complutense de Madrid

Translated by Jody Doran

World Scientific
Singapore • New Jersey • London • Hong Kong

Published by
World Scientific Publishing Co. Pte. Ltd.
P O Box 128, Farrer Road, Singapore 912805
USA office: Suite 1B, 1060 Main Street, River Edge, NJ 07661
UK office: 57 Shelton Street, Covent Garden, London WC2H 9HE

British Library Cataloguing-in-Publication Data
A catalogue record for this book is available from the British Library.

THE COUNTINGBURY TALES: FUN WITH MATHEMATICS

Copyright © 2000 by World Scientific Publishing Co. Pte. Ltd.

All rights reserved. This book, or parts thereof, may not be reproduced in any form or by any means, electronic or mechanical, including photocopying, recording or any information storage and retrieval system now known or to be invented, without written permission from the Publisher.

For photocopying of material in this volume, please pay a copying fee through the Copyright Clearance Center, Inc., 222 Rosewood Drive, Danvers, MA 01923, USA. In this case permission to photocopy is not required from the publisher.

ISBN 981-02-4032-5
ISBN 981-02-4033-3 (pbk)

Printed in Singapore by Regal Press (S) Pte. Ltd.

Table of Contents

Prologue ix

The Mathematics of a Sandwich 1

Nim 15

The Bridges of Königsberg 31

"Solitaire" Confinement 41

The Mathematician as a Naturalist 55

Four Colors Are Enough 69

Leap Frog 89

Abridged Chess 105

The Secret of the Oval Room 113

Bibliography 121

Dedicated to Miguel and Mayte, with whom I have spent so much time playing.

Prologue

One day, an idle man from a city in Germany, Königsberg, came up with an odd, useless question, the only point of which seemed to be based on the difficulty involved in answering it: could he plan a route that would cross the seven bridges across the River Pregel that joined the various quarters of the city with the island located in the middle? The question traveled by word of mouth and through the heads of many without an answer until it alighted upon the head of Euler. There it nested, and, following a period of incubation, one of the most important branches of mathematics, topology, was hatched.

A French gentleman and compulsive gambler had an important problem. What, in any given situation, would be the right bet in order for the gamblers to have the same odds? Fortunately, the gentleman also had a friend, who was none other than Blaise Pascal. The question nested in Pascal's head, and in an exchange of letters with his friend Fermat based on the gentleman's question, he created the theory of probability, which has given birth to an endless number of disciplines in mathematics, physics, etc.

Games and beauty are found in the origin of a major part of mathematics. If the mathematicians from throughout history have had such a good time playing and contemplating their games and their science, why not try to learn it and pass it on through games and beauty?

This is the fundamental idea that underlies the stories and games I am presenting here. My wish would be for them to enable many people to find the pleasure and satisfaction I myself have found in them, and for this to be like a bridge to find the same pleasure in other mathematical endeavors that may look more serious and complicated, but if we look carefully, display basically the same fun, playful spirit.

The Mathematics of a Sandwich

You're holding in your hand a tasty summer sausage sandwich. Open it up. Take a look at one of its appetizing slices. What shape does it have? Of course it depends. If you were given little pieces, they were probably sliced thin and in circles. But if you cut it yourself generously and wanted good, hearty slices, you probably cut the sausage on an angle, more or less like this:

What is the shape of your slice? It looks like an ellipse, right? Could it be an ellipse? What did you do to make it an ellipse? What is an ellipse? Remember: an ellipse is a set in the plane made up of all the points whose sum of the distances to two fixed points, the foci, is constant. It would be like a miracle if you, who only wanted to cut yourself a good slice of summer sausage, managed to form an ellipse. It seems like you'd have to put a lot of care into making sure that the slice would come out in that shape. And yet, there you have it: an ellipse.

In order to really see it properly, imagine the time when the sausage was still intact, just a minute ago, and imagine the cut you made in it just now. Imagine that, tangent to your knife, on either side, and tangent to the inside wall of the sausage, you have placed (can you imagine?!) two ping-pong balls that just happen to fit perfectly. So:

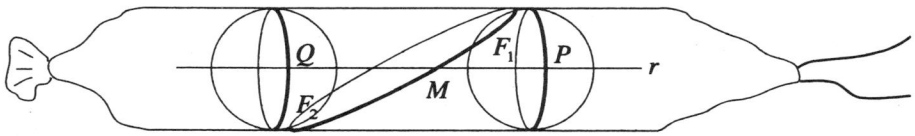

Now consider points F_1 and F_2 in which the ping-pong balls are tangent to the cut, and watch what happens with any point from the edge of the cut. Draw a line parallel to the axis of the sausage. This line will be tangent to the ping-pong balls at P and Q. Meanwhile, MF_1 is tangent to the ball on the right and MF_2 to the one on the left. If you draw any two tangents from an outside point to a ball, the segments determined by the point and the tangent points will be the same length.

Thus it turns out here that $MF_1 = MP$, $MF_2 = MQ$, $MF_1 + MF_2 = MP + MQ = PQ =$ constant, because PQ is the same length for any r we could draw on the surface of the sausage parallel to the axis. Thus, point M follows the path of an ellipse with foci F_1 and F_2.

So we see that an ellipse is very easy to draw. You take a slice of sausage, place it on a piece of paper and the edge of the grease spot made on the paper is an ellipse. If you want a less greasy ellipse you can use the tube from a roll of toilet paper or paper towels: cut it on an angle with a knife, place the edge on a piece of paper and trace it with a pencil.

You would think that an ellipse, as defined ("the geometric location of the points of the plane such that...") would be merely an invention by mathematicians. And yet this is not the case. Ellipses turn up everywhere. You might even find them in your soup! Tilt the soup pot a bit. The edge of the surface is an ellipse. Have you got an adjustable reading lamp? Turn it on, put it on top of your desk and take a look at the shape the illuminated area of your desk is when you tilt the lamp shade. Doesn't it remind you of the slice of sausage? Could it also be an ellipse? Of course, and for the same reason! Look closely. The area of light is more or less that of a cone with a circular base.

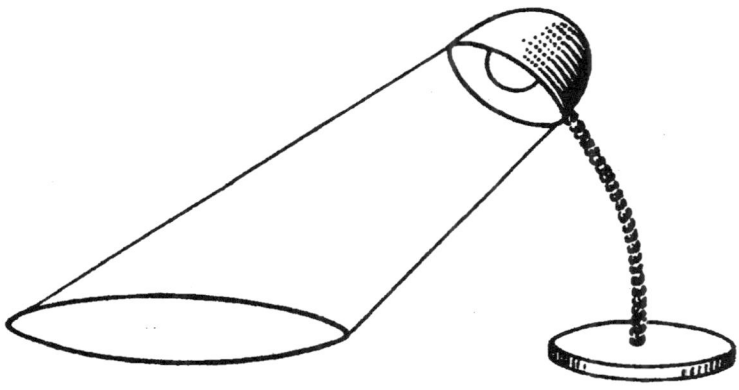

Cutting the illuminated area with the desk is like cutting an enormous conical sausage with the desk. What a crazy idea! Let's use the same trick as before to see whether the figure is an ellipse, all right? We'll put two balls, this time different sizes, tangent to the cone and to the plane of the desk on either side of it, as you can see in the figure. Just like before, it turns out that $MP = MF_1$, $MQ = MF_2$, and thus $MF_1 + MF_2 = MP + MQ = PQ =$ constant independent of r. So M describes an ellipse with foci F_1, F_2.

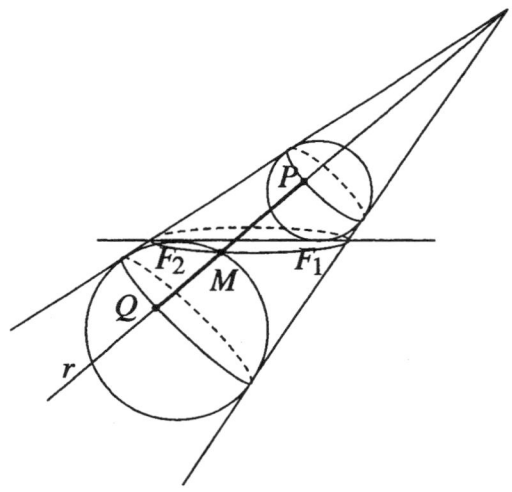

Ellipses appear quite unexpectedly on many other occasions, such as this one. You have a ladder leaning against the wall. You're climbing up the ladder. You still haven't reached the top. All of a sudden, oh no! It begins to slip. By the time the accident ends up with you splayed across the floor, you probably haven't had time to realize it, but did you know that your foot, for example, has drawn a clean segment

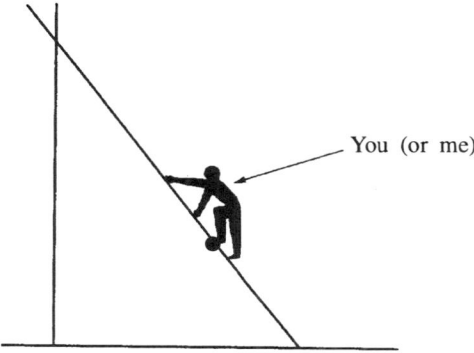

of an ellipse in the air? Imagine! The sum of the distances from your foot to two set points was constant the whole time of your fall (assuming, of course, that you didn't make a weird leap into the air when you realized what was happening, which would probably not have been a bad idea). Who'd have thought of it!

Just by knowing a little analytic geometry, you can see it easily.

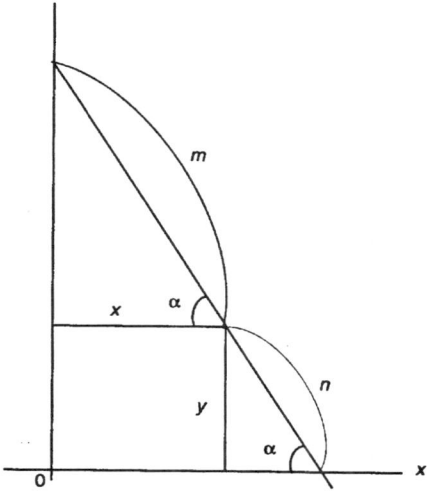

Suppose your foot is at the large dot shown on the falling ladder (m and n, fixed; angle α, getting smaller). The idea is to figure out the type of curve that describes the movement of the large dot on the ladder. Its coordinates: $x = m \cos \alpha$, $y = n \sin \alpha$.

Thus, since $\cos^2 \alpha + \sin^2 \alpha = 1 = \dfrac{x^2}{m^2} + \dfrac{y^2}{n^2}$

it turns out that your curve is: $\boxed{\dfrac{x^2}{m^2} + \dfrac{y^2}{n^2} = 1}$

Proving that this is an ellipse represented by the following figure with foci F_1, F_2 is done as follows:

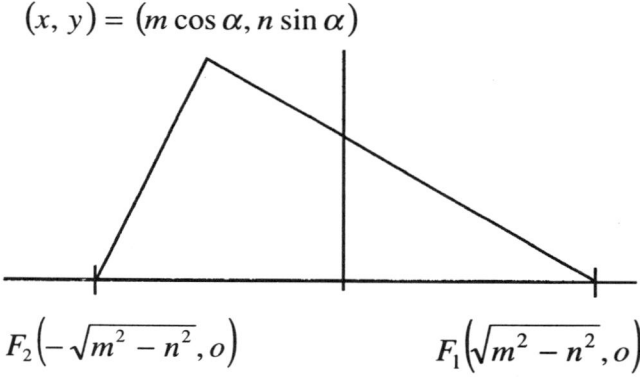

The sum of the distances from the point to F_1 and F_2 is

$$d_1 + d_2 = \sqrt{\left(m \cos \alpha - \sqrt{m^2 - n^2}\right)^2 + n^2 \sin^2 \alpha} +$$

$$\sqrt{\left(m \cos \alpha + \sqrt{m^2 - n^2}\right)^2 + n^2 \sin^2 \alpha}$$

$$= \sqrt{m^2 \cos^2 \alpha + m^2 - n^2 - 2m \cos \alpha \sqrt{m^2 - n^2} + n^2 \sin^2 \alpha} +$$

$$\sqrt{m^2 \cos^2 \alpha + m^2 - n^2 + 2m \cos \alpha \sqrt{m^2 - n^2} + n^2 \sin^2 \alpha}$$

$$= \sqrt{m^2 + \cos^2 \alpha (m^2 - n^2) - 2m \cos \alpha \sqrt{m^2 - n^2}} +$$

$$\sqrt{m^2 + \cos^2 \alpha (m^2 - n^2) + 2m \cos \alpha \sqrt{m^2 - n^2}}$$

$$= \left(m - \cos \alpha \sqrt{m^2 - n^2}\right) + m + \cos \alpha \sqrt{m^2 - n^2} = 2m$$

In other words, the sum of the distances from the point to F_1 and F_2 is constant, $2m$. This is the major axis of the ellipse. So *the major semiaxis of the ellipse is precisely what was left to get to the top of the ladder.*

The minor semiaxis is the distance on the ladder you had climbed when it began to slip, n.

Here is a simple way of making an ellipse emerge based on its definition. Take a piece of thread and, using a thumbtack, attach each end to a point on a piece of paper such that the thread is a bit loose between the ends. Now take a pencil and use the point to pull the thread taut. Now let the pencil point slide across the paper along the taut thread. The figure is clearly an ellipse, since at any given time the sum of the distances from the point you've drawn to the two thumbtacks, the foci, is fixed and equal to the length of the thread.

One more unexpected way of obtaining an ellipse is the following: Take a piece of paper. Using a compass, draw a circle. Cut it out. Mark any point *P* within the circle that is not the center. Now take

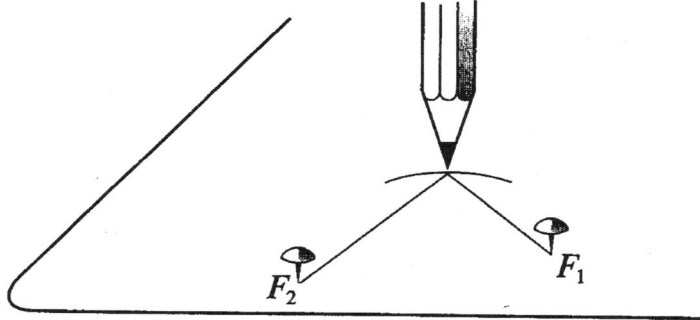

the piece of paper and fold it such that the edge of the circle touches point *P*. Unfold and fold again in such a way as to meet the same condition. Make a lot of similar folds. When you finish, you'll see that all the folds envelop a figure that looks like an ellipse. Could it be an ellipse?

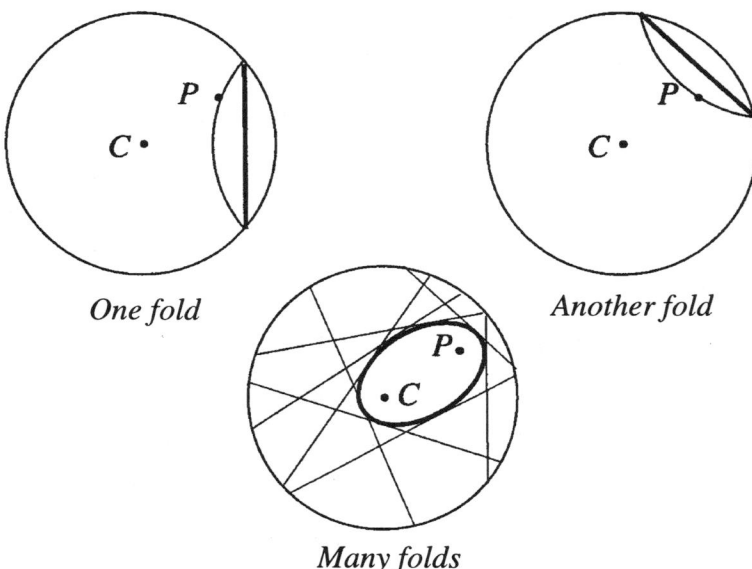

One fold *Another fold*

Many folds

Yes! And you'll see that it's not difficult to prove. Consider any fold such as the one in the following figure.

The arc that passes through P is symmetrical to arc AQB with respect to the fold. Draw the perpendicular line from P to the fold (AB). AB is the perpendicular bisector to PS. Connect C to S, cutting AB at M. Thus we have $MP = MS$ and so $MC + MP = MC + MS$ = radius of the circle. Thus we find that M is on an ellipse with foci C and P with the major axis of a length equal to the radius of the circle. Furthermore, for any other point T on AB we have $TP + TC = TS + TC > CS$ (since side CS of the triangle TSC is less than the sum of the other two sides). So T is not in the ellipse. In other words, *the only point on AB in the ellipse is M, which is to say that AB is tangent to the ellipse at M*. This explains why the folds envelop the ellipse.

Look at the figure and you'll also see that because of the symmetry, angle PMA is equal to AMS, which is equal to CMB, which is to say that *it turns out that the tangent AB to the ellipse with tangent point M is the outside bisector of the angle whose vertex is M and whose sides are the segments that go from M to the foci of the ellipse*. Remember this fact, because it will be very useful later on.

Another property, which you can deduce by looking closely at the previous figure, is also very interesting. The projection of P on the tangents to the ellipse follows a circle that is homothetic to the original circle with a homothetic center at P and a homothetic ratio of 1/2.

10 The Countingbury Tales: Fun with Mathematics

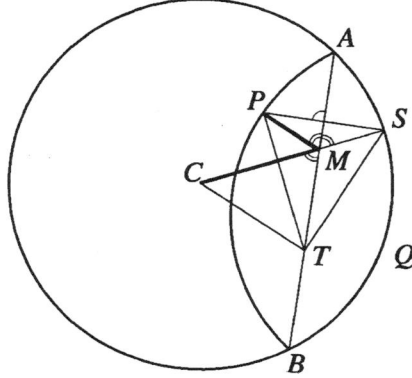

Here are a few questions for you to think over:

Can you think of a method like the thumbtack one for drawing a parabola or hyperbola?

In the folding method, what could you do to obtain a hyperbola from a similar method using folds?

To finish this elliptical snack, which began with a sausage sandwich, I'm going to offer you a dessert you won't want to pass up. Look around and see if you can find one of those round metal cannisters which they sell butter cookies in. Measure the diameter of its interior, and, using construction paper, draw and cut out a circle whose diameter is equal to the radius of the inside of the cookie cannister. Line the bottom of the cannister with a piece of regular or construction paper, where the miracle of the ellipse will appear. Now the idea is to place the small circle on the lined bottom of the cannister and move it along with its edge touching the inside wall of the cannister without slipping; in other words, you want to *make it turn in a circle without slipping*.

To do this, the best thing is to line the inside wall of the cannister with, for instance, a strip of double adhesive tape like you use to stick carpeting to the floor. Make a hole at one point in your small circle, the one that moves, stick your pencil tip through it, and you'll see that, when you move it like we've described here, without letting it slip, the pencil will map out something like an ellipse on the lining paper. Is it an ellipse? I'll leave it for you to check.

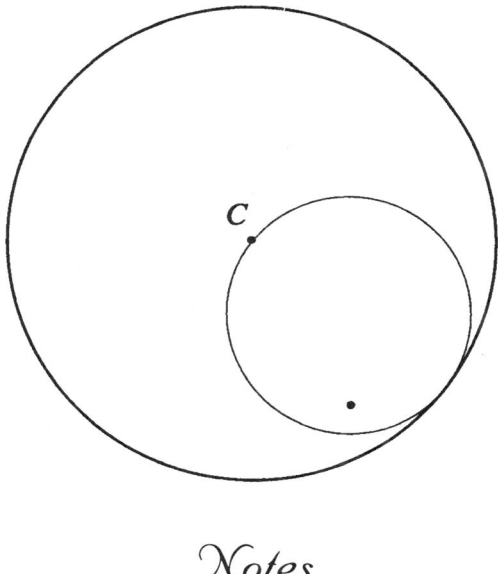

Notes

The geometry of cones, ellipses, hyperbolas and parabolas is one of the deep areas of mathematics that, due to its beauty, harmony and richness of ideas, was one of the earliest to be developed.

The three great geniuses of Greek mathematics, Euclid, Archimedes and Apollonius, are responsible for the fact that by the third century

BC almost as much was known about the properties of cones as we know today.

All of this evolution took place around the great center of learning of Alexandria in Egypt, the city founded in 322 BC by Alexander the Great that served for more than five centuries, until the middle of the third century AD, as mankind's most important cultural center, with its great museum, its more than 500,000-volume library and its schools of the different branches of knowledge. The school of mathematics was founded by Euclid around 300 BC in the time of Ptolemy I. Only a few curious anecdotes are known about the life of Euclid. Once a pupil interrupted his explanation of geometry to ask him what was to be gained with such strange cogitations. Euclid addressed his slave, who was in attendance: "Give him a coin and be gone with him. He is one of those who only wishes to know for gain." Euclid's principal work, the *Elements*, has been one of the most influential in the thinking of humanity. It was used as a textbook throughout the world until well into the nineteenth century. It is said that only the Bible surpasses it in number of editions. Its importance lies in the fact that it was the model for the method of thought from which generations of the western world have learned for more than 22 centuries.

Archimedes (287–212) is considered the most brilliant mathematician of Antiquity. He was born in Syracuse (Sicily), a Greek colony. He was the son of Phidias, an astronomer, somehow related to King Hieron II of Syracuse. Archimedes studied in Alexandria, assimilated and perfected the methods of Euclid's disciples and returned to Syracuse, where he surprised the world with his many discoveries and inventions, among which was his famous principle "Any body submerged in a liquid…" This principle is related to the discovery of the fraud committed by a goldsmith against King Hieron. The latter had commissioned the making of a gold crown, and entrusted him with a certain amount of pure gold. The king suspected that the goldsmith had removed some of the gold, replacing it with the same weight of silver. Archimedes, in one of the city's public baths, was meditating on how to ascertain whether this was true. Suddenly, on feeling his body floating in the water, his mind was struck by a flash. An amount of gold of a given weight has a smaller volume than the same weight in silver. Therefore, if the crown contained silver as the king suspected, its volume would be greater than that of the gold the king gave to the goldsmith. All it would take would be to place the crown in the water, measure the displaced water, place an amount of gold with the same weight as the crown in the water, measure the displaced water, and compare. Archimedes couldn't contain himself. He left the baths and ran through the streets of Syracuse naked, shouting, "Eureka, Eureka!" (I found it, I found it!)

And yet, despite all of his famous inventions (lever, parabolic mirrors to burn the Roman ships that surrounded Syracuse from a distance, a sort of giant screw to conveniently draw water from the well…), what Archimedes was most esteemed for were his geometric discoveries. He ordered a sphere inscribed in a cylinder engraved on his tombstone as a reminder of his discovery that the volume of a sphere is two-thirds the volume of a cylinder.

But the one who knew the most about conics was Apollonius of Perga. He was born around 262 BC in Perga, a Greek city located in what is now Turkey. He also studied and taught in Alexandria, where he died in 190 BC. He also worked in optics and astronomy, introducing very original methods and obtaining very profound results.

Archimedes' screw. An ingenious method for drawing water from a well.

Nim

The game I'm going to explain to you now is quite old and very interesting. *Nim* in Old English means "remove" or "take away." It became fashionable in Europe thanks to a celebrated film, *Last Year at Marienbad*, in which the hero seems to become dazzled by the game and uses it to kill time at the Marienbad spa in Czechoslovakia, renowned throughout Europe since the sixteenth century.

This is a game for two players, *A* and *B*. You set up four rows of pebbles (the Marienbad guy used matches), one row with 1, another with 3, another with 5 and the last with 7 pebbles. We are going to start playing with three rows of pebbles as shown below. Later we'll see that our strategy is the same for playing with any number of rows and pebbles per row.

		O	O	O	Row 1
	O	O	O	O	Row 2
O	O	O	O	O	Row 3

Player A may remove one or more pebbles from any row he chooses. For example, he removes 2 from row 2, leaving the following arrangement:

```
                    O    O    O      Row 1
                         O    O      Row 2
     O    O    O    O    O           Row 3
```

Now it's B's turn; she may remove as many pebbles as she wants (always one or more) from whatever row she chooses. For example, she removes 2 stones from row 1, leaving the following arrangement:

```
                              O      Row 1
                         O    O      Row 2
     O    O    O    O    O           Row 3
```

Then it's A's turn, etc. The player who takes away the last pebble is the winner.

If you have a friend or enemy somewhere close by, invite him or her to play for a while. The game is simple to learn and it is not at all easy to guess the right strategy to win all the time.

When you've played for a good while and given a little thought as to how you should play to win, come back to the book, by yourself if possible. Don't tell your friend that someone is going to explain the strategy for winning, because among other reasons, what I'm going to teach you will allow you to almost always beat a person who doesn't know the strategy, and it will let you beat your opponent every time, even if he does know it, as long as you make the first move. See you later.

The strategy I'm about to explain to you is based on the binary numbering system. Don't let that scare you. You'll see how easy it is to put a pile of pebbles into the binary system without writing a thing, without even knowing how many pebbles there are in the pile.

Here are the pebbles you have:

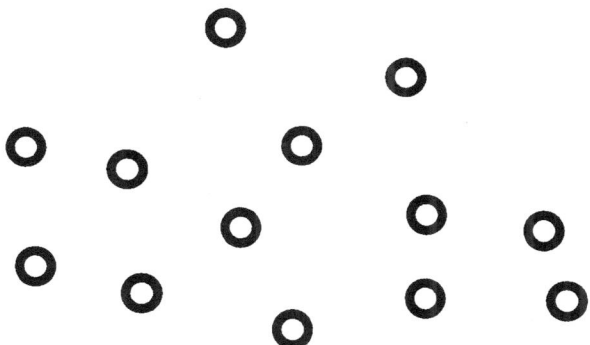

Line them up, like this:

You don't have to count them. Now, starting on the left, take the first one and put it underneath the second one, take the third and put it under the fourth, etc. Like this (no need to count):

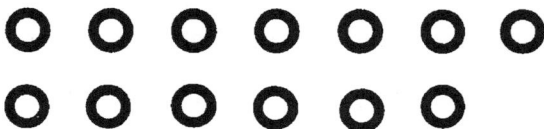

Now, starting on the left, take the first pair and put it under the second pair, and take the third pair and put it under the fourth, etc., like so:

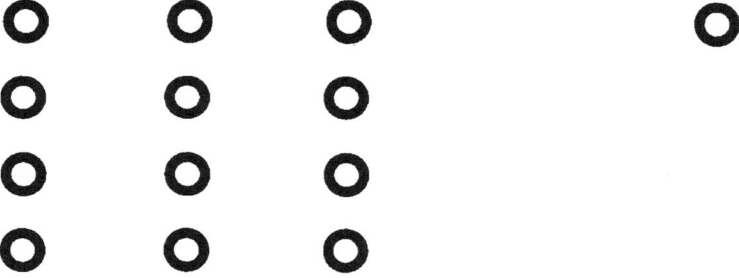

Now take the first pile of 4 pebbles and put it under the second pile, the third under the fourth (at this point, since there are no piles of four pebbles left, we'll stop).

This is how it looks now:

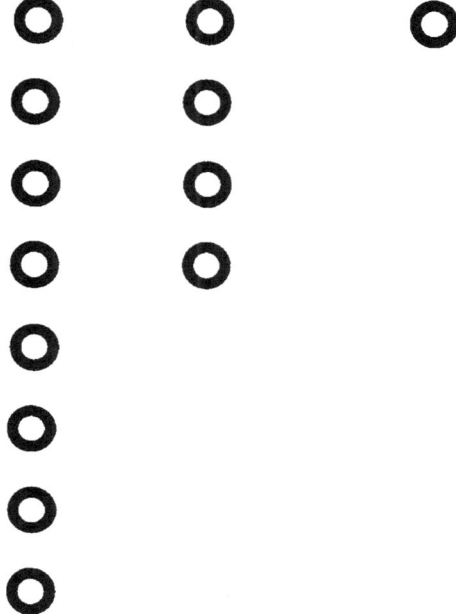

Now you take the first pile of 8 pebbles and put it under the second pile (but since there is no second pile, we won't do anything; we're finished).

I think you get the idea of what we've done here. We've distributed the stones into groups of 1, 2, 4, 8, 16,..., pebbles. In our case we end up with 1 group of 8, 1 of 4, 0 of 2, 1 of 1, which is to say that in the binary system the number of pebbles is 1101. Note that we haven't had to write a thing or do any division at all, or even count the pebbles.

Try this example with me for practice.

20   *The Countingbury Tales: Fun with Mathematics*

```
    o o o       o
   o     o o o o   o
    o o o   o
```

Row 1   o o o o o o o o o o o

Row 2   o   o   o   o   o o
        o   o   o   o   o

Row 3   o   o   o o
              o   o   o
              o   o
              o   o

Row 4   o   o o
                          o   o
                          o
                          o
                          o
                          o
                          o
                          o

The number of stones in the binary system is 1011.

This is just the preparation. Let's get to the game. Suppose you're the first one to play, so you're *A*. Your opponent is *B*. Place the stones of each row in the ordered binary system, that is, in such a way that the eight-stone piles from each row are in a single column, the four-stone piles in another, the piles of 2 in another and the piles of 1 in another. Like this:

○○○○		○	Row 1
○○○○			Row 2
	○○	○	Row 3

Now take a look at this arrangement of the stones by columns. In the first column on the left there are two groups of 4 pebbles, in the second column, one group of 2, and in the third, two groups of 1. Your objective in playing is to leave an even number of groups. So in this case you can remove the group of 2 from row 3, leaving *B* with the following playing situation:

22 *The Countingbury Tales: Fun with Mathematics*

O O O O O		O	Row 1
O O O O			Row 2
		O	Row 3

Now it's up to B, who will remove as many stones as she likes from whatever row she feels like. Suppose, for example, that she takes 2 from row 1, leaving you with the following situation:

		O	O	O	Row 1
	O	O	O	O	Row 2
				O	Row 3

Now, go with your strategy. Place the stones in each row into the binary system. Like so:

	oo / o	o	Row 1
oooo			Row 2
		o	Row 3

To continue with your strategy of leaving all the columns with an even number of groups, note that you can remove 2 stones from row 2, thus leaving B with the following situation, once you've put the stones of each row into the binary system:

	oo / o	o	Row 1
	oo / o		Row 2
		o	Row 3

Suppose that *B* takes a stone from row 2, leaving you with this:

	O	O	O	Row 1
			O	Row 2
			O	Row 3

So you put them into the binary system, like always:

	O O O	O	Row 1
		O	Row 2
		O	Row 3

Note that if you take away the 3 pebbles from row 1, *B* is left with:

			Row 1
		O	Row 2
		O	Row 3

a situation that leaves her with no alternative but to let you win.

Take a good look. It doesn't matter at all whether the game starts with 3, 4 or 5 pebbles. Nor does it matter that there are only three rows. The one thing that's important is for you to be able to go ahead with your strategy: placing the stones of each row into an ordered binary system and removing as many stones as needed from whatever row necessary so that, when the stones are put in the binary system, there is an even number of piles in all the columns.

Let's suppose that you begin playing with four rows of 14, 11, 9 and 7, respectively, and you're A, meaning that you go first. The initial situation in the binary system is:

26 The Countingbury Tales: Fun with Mathematics

○○○○○○○○○	○○○○	○○		Row 1
○○○○○○○○		○○	○	Row 2
○○○○○○○○○			○	Row 3
	○○○○	○○	○	Row 4

The 8-pebble column has 3, the 4-pebble column has 2, the one with piles of 2 has 3 and the one with piles of 1 has 3. In order to follow your strategy you have to look at the first column with an odd number of groups, in this case the first on the left, which has three. In each of these three groups there are eight stones. Note that $8 = (4 + 2 + 1) + 1$. So, by moving or removing stones from any of these three groups of eight you can fill the other sections of the same row or leave them empty if necessary so that after moving the stones of one single row, all the columns have an even number of groups of stones. Here, for example, you pick up the eight stones from the group of eight in the first row, take one of the stones from the group to put in the one-stone section and remove the other seven, and also remove the two from the two-stone section of the first row. Thus the first row looks like this in the binary system.

	o o o o		o	Row 1

This way each column has an even number of piles.

Note the following two important facts:

1) You can always carry out your strategy *provided that there is a column with an odd number of piles*. If this is not the case and all the columns have an even number of piles, don't sweat it. Playing

by the rules, taking one or more stones from a single row, you'll always leave *B* some column with an odd number of piles and, if she knows your strategy, she'll beat you if she is also using it. But if she doesn't know it, it's almost a sure thing she'll leave a good set-up for you to use in a later move.

2) Often when it's your turn you've got various possibilities for putting your strategy into practice, *but as long as there is a column with an odd number of piles, by taking away some of the stones the way I've shown you, you'll always be able to use your strategy.* For example, in the case on page 28, you could remove all the pebbles from row 2 and you'd reach your objective. The other play explained there uses the longer row.

If this is getting to be a drag, take a break and come back another day. If you're only interested in winning as many times as possible, don't come back here, because what's coming up next is the mathematical explanation of why the strategy I've given you works.

The fact that the strategy always leads to victory is fairly clear. If *A* can always leave *B* with a situation where there is an even number of piles in each column in the binary system, it is either because this even number is zero for each column (in which case *A* has taken all the pebbles in his last move and already won) or because there are at least two rows with stones, and so *B* cannot win from this position (she must take stones only from one single row).

What requires a more detailed explanation is the fact that *A* may always follow his strategy if at some point he was able to start it, that is, if at some moment, either at the beginning or at some other point in the game, he received the stones in such a way that, placed in the binary system, there was an odd number of piles in at least one of the columns. The reason for this fact is easy to see considering the very recipe I gave earlier. Player A should look at the first column on the left with an odd number of groups. In each there are 2^p stones; for example $32 = 2^5$ stones. Note that

$$2^p = (2^{p-1} + 2^{p-2} + \cdots + 2^1 + 2^0) + 1$$

for example

$$32 = 2^5 = (2^4 + 2^3 + 2^2 + 2^1 + 2^0) + 1 =$$
$$= (16 + 8 + 4 + 2 + 1) + 1$$

So in any given group there are enough stones to fill or empty the next sections of the same row in such a way that all the columns end up with an even number of groups of stones.

Now you might think the following. Suppose we are playing with twelve pebbles and three rows and start by randomly placing the stones in three rows. From the beginning, what are the odds, if I'm the first to move, that the situation will be in my favor and allow me to apply the strategy so that, even if my opponent knows what it is, I can beat him? Try to calculate it. In almost every case you'll be able to win. Of the twelve possible distributions there is only one in which the odds will be against you, which is 2, 4, 6. All the others are in your favor.

Nim has been programmed with this strategy so that computers can play…and win. But don't get scared if you have to play against one of them. Tell it you accept, but that you want to play with 25 rows. The computer will probably be a bit daunted, since they're usually not programmed for more than 10 rows. It's good for computers to know that there are some things they have yet to learn...

Notes

There are a lot of games in the same vein as the one we've just studied; maybe you've encountered some playing at the video arcade. A really old one for two players consists of starting with a pile of stones. The first player, A, is allowed to remove 1, 2 or 3 stones. Then it is B's turn, and he may also remove 1, 2 or 3 stones. The winner is the one to remove the last stone. The strategy for this game is very simple. If the two players know it and follow it, then you can tell from the beginning who is going to win, depending on the original number of stones. Think about it to see if you can figure it out.

Nim, just like this other game, allows simple variations, such as the loser being the one to remove the last stone. How would the strategy vary then? You can also think about combining the two games. Start with several piles of stones, three for example, one with 3, another with 4 and the other with 5 stones. You can remove 1 or 2 stones from a single pile. Are you up to the task of finding a strategy?

The Bridges of Königsberg

One of the important branches of modern mathematics, topology, was born with the following riddle that the great Euler described and solved in one of his articles: "The problem that, as I understand it, is very well known, is formulated as follows: In the city of Königsberg, in Prussia, there is an island called Kneiphof, around which flow the two branches of the Pregel River. There are seven bridges, *A*, *B*, *C*, *D*, *E*, *F*, and *G*, that cross the two branches of the river (see figure below). The question is a matter of determining whether it is possible for a person to take a walk in such a way as to cross each of the bridges one single time. I have been told that while some denied that it was possible and others doubted it, no one maintained that it was really possible."

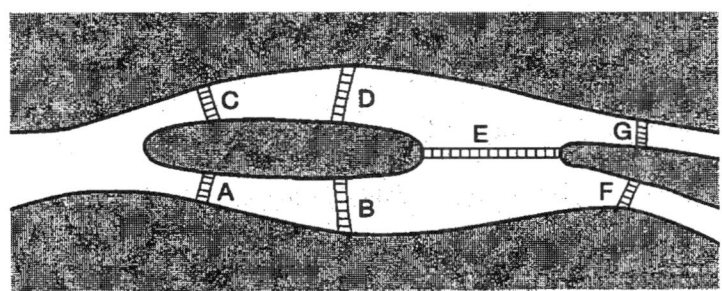

Where do you start to tackle this problem? Think and observe. There are many aspects of the problem that are totally irrelevant and are not the least bit important. For example, whether the island is big or small, whether the bridges are narrow or wide, straight or curved, long or short. What is essential is the diagram, what it is the bridges join and how these meeting points interact. *What is essential, then, is the following: Is it possible to draw this path without lifting pencil from paper and without repeating any line?*

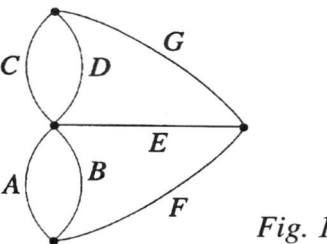

Fig. 1

This may bring back memories from when you were little. Are you able to trace the following figures without lifting pencil from paper and without repeating any line? Could you do the same thing starting and ending at the same point?

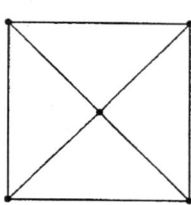

Fig. 2

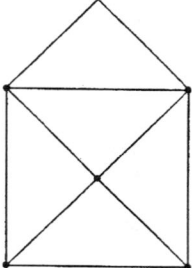

Fig. 3

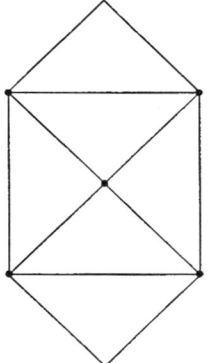
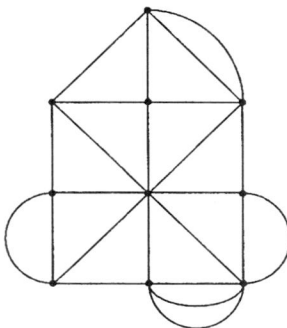

 Fig. 4   Fig. 5

Try and try again. You're almost sure to be familiar with Figure 3 and have probably done it many times, but you'll have a hard time finishing where you started. Figure 4 is so easy to trace that unless you try to do it wrong, you can start at any point and return to the same point without repeating any arcs, covering the entire path almost effortlessly. Figure 2 looks easier, with fewer segments, but it, just like Figure 6 below, which just has three segments, is clearly

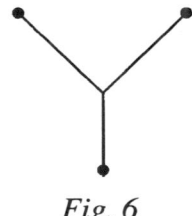

Fig. 6

impossible ("trivial," as some people would say almost insultingly). It looks like Figure 5 would be impossible for any human (or non-human, for that matter) to analyze, but here is a solution:

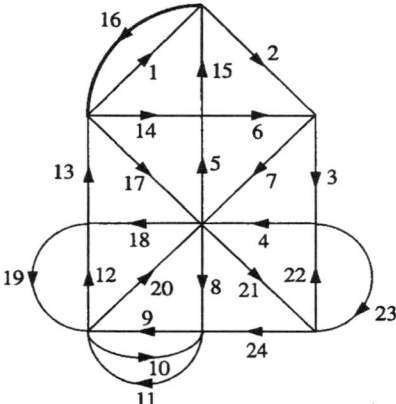

Fig. 7

What is the mystery of the arcs for each of these cases? How can you find out whether it's possible to draw a given figure as asked? And if it is possible, how do you find the solution?

Let's start with simple cases:

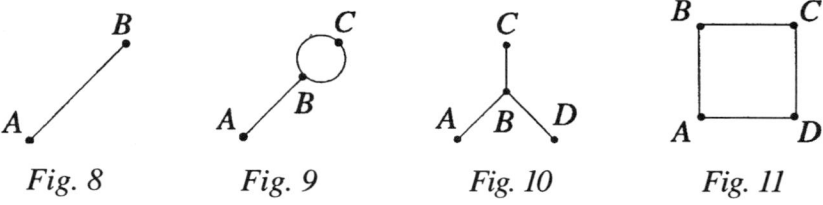

Fig. 8   Fig. 9   Fig. 10   Fig. 11

Figure 8 can be traced (let's hope so!), but you can't start and end at the same point. Figure 9 can be done starting *A* and ending at *B*, and also starting at *B* and ending at *A*, but if we start at *C* we can't do it. Figure 10 cannot be completed at all. Figure 11 can be traced, starting at any point and ending at the same point. What distinguishes the vertices is obvious: the number of possible ways in and out of each of them, or the number of arcs that meet at each. Here are the numbers, the valence of each vertex:

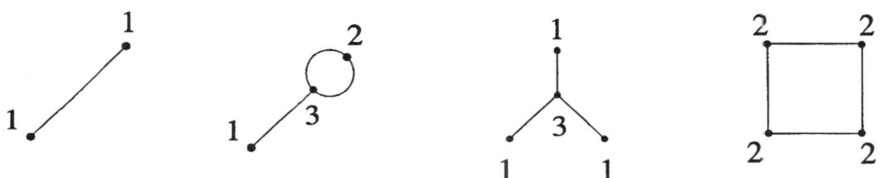

Why is this number so important? Hey, what we're looking for are ways in and ways out. What gets us stuck at a vertex is when there's no way out, right? Sure, but having lots of ways in and ways out is not always a good thing. Figure 12 has more ways in and ways

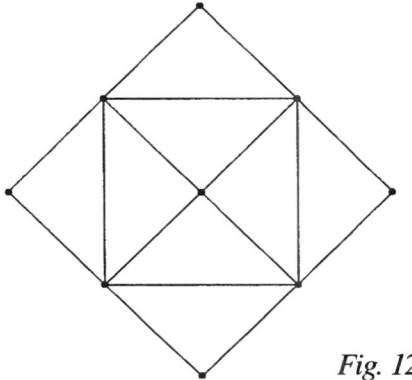

Fig. 12

out than Figure 3, and yet Figure 3 is possible and Figure 12 impossible. Let's think of a figure that can be traced with the same starting and ending point. For each vertex of the path, since we don't stop there, it turns out that we enter as many times as we leave, naturally along different arcs.

Thus each vertex has an even valence. The first one does, too, since that's where we stop.

Thus, if it's possible to trace a figure and end up at the same vertex as the starting point, all of its vertices must have an even valence.

Does this completely clarify our problem? Not yet! Is it true that if all the vertices have an even valence, we can follow the path and end up at the starting point? Let's take a look. What we know for certain is that we'll never get stuck on our path unless it is at the starting vertex S, because since each vertex has an even valence, when we enter one other than S for the first time, we are left with an odd number of starting arcs, that is, at least one arc; when we enter for the second time we are again left with an odd number, since we have used three arcs that meet at that vertex; so whenever we enter, we are able to leave. Therefore, if we just follow our figure any old way starting at S, we'll only get stuck when we reach S once again. If we've walked around the entire figure, our problem is solved. But what if we haven't? If on our path, C, we still have arcs to trace, what we can do is to expand our path and make a bigger one to continue checking the rules of the game. When we reach the first vertex S_1 where untraveled arcs start, let's follow them. As before, we can't stop unless it's at S_1. Now, when all the arcs that start at S_1 have been followed, we continue from S_1 along the initial path, C, until we reach the first vertex, S_2, where there are untraveled arcs either from path *C* or in the extension we just made. This way we end up traveling along all the arcs.

Thus, *if all the arcs have an even valence, it's possible to follow the path of the proposed figure and, what's more, we have the solution for tracing our path. And this solution tells us that the starting vertex, which may be any vertex, is necessarily the same as the ending vertex!*

But what if there are odd vertices? Take a look at Figure 9. If you leave from *A* or *B* you can do it, but if you leave from *C* you can't. Vertices *A* and *B* are odd, and *C* is even. What's the mystery? Whatever it was that led us to a solution before might tell us how to proceed now. On a path like the one we have to follow here, there is a starting vertex, an ending vertex and all the intermediate ones along the way. But an *intermediate vertex (neither a starting one nor an ending one) has the same number of ingoing arcs as outgoing arcs; in other words, it has an even valence. So if a figure allows a path like the one we are looking for here, every intermediate vertex must be even.* But all but two of the vertices are intermediate. Therefore, if a figure has more than two odd vertices, it is impossible. On the other hand, if a figure has two odd vertices, clearly if we try to trace it by the rules we will have to start at one of the odd vertices and try to finish at the other odd vertex.

There's just one more question we must answer before having our problem completely solved. If a figure has two or just one odd vertex, can it be done? And what's the solution? What we have learned up to now can clarify the matter. If a figure has one or two odd vertices, we start decisively at one of them, *S*. We can't get stuck at any even vertex, since if we enter we can always exit, and we can't get stuck at S, since when we start we use up one of its arcs, which leaves an even number of them, meaning that if we enter again we can leave. Since we finally do get stuck (as there are only a finite number of arcs) it's obvious that we got stuck at the other odd vertex, which demonstrates that *there cannot be just one odd vertex*. Now our question is: with path C have we covered the entire figure? If we have, congratulations, our problem is solved. If not, let's proceed as before. We leave S along path C until reaching the first vertex S_1 from which untraveled arcs in path C branch out. Note that all the vertices that still have untraveled arcs in C are missing an even number of arcs to be covered. Thus, leaving from S_1 along arcs not in C, we don't get stuck at any vertex other than S_1. So we end up at S_1 by way of path C_1 of arcs not in C, having covered all the arcs of S_1 that were not in C. Now we can continue along C until reaching the first vertex, S_2, with arcs not in either C or C_1. We follow the same procedure, and thus end up covering all the arcs of the figure.

You can practice this method by coming up with potentially complicated figures and challenging a friend to trace them. For instance, the following:

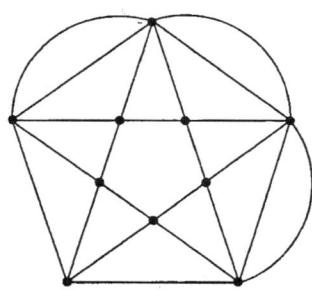
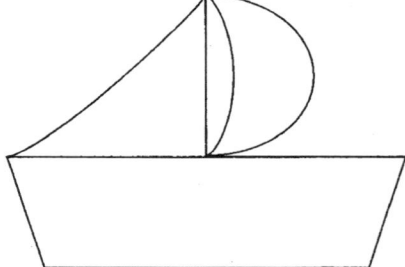

You can also imagine that you have an airplane in Königsberg that allows you to make a single leap from one point to any other point you wish. Then could you cover the path? What is the basic condition for a figure in order to be able to cover it in a single leap?

Notes

The method we have seen for solving problems like the Königsberg bridge also enables us to solve the problem of reaching the center of any maze we're given, even without knowing its structure at all. Let's set a goal of getting from the starting point to the spot where the treasure is located in the maze in the garden of R. Ball, one of the greatest mathematical game writers of all times. Suppose that we don't have a map, and so our aim is to cover the *entire* maze (we assume, of course, that the treasure is in clear view at some accessible point in the maze) and to exit through the only entrance it has. Can we do it?

Yes! The path that defines the maze (the italicized line in the second figure) has arcs and bifurcation points. Since we want to travel along each arc once each way, we repeat each arc twice. Once we have done this, we clearly have a figure like the ones we have been studying in this chapter with all even vertices. This way we can travel along all of it without repeating arcs, starting at any point. Furthermore, we can do it without knowing the layout of the maze. The only thing we need is to be able to mark the arcs we have already covered by some means.

The maze of the garden of Rouse Ball, and its treasure.

To do this, all we need is a piece of chalk; at each fork we mark which path we have taken with an arrow to keep from taking the same path when we return to the same point.

Naturally, this procedure doesn't give us the shortest route for reaching the treasure, but it does ensure that we'll reach it and be able to get out once again.

"Solitaire" Confinement

There's no question that the best thing for getting a person who's in solitary confinement out of it is to put him in a group. But we're not going to worry about that meaning of solitary or that type of group.

In the 18th century a French aristocrat was put into solitary confinement in the Bastille. The solitary nobleman invented a game of solitaire to make his solitude more bearable. The game became all the rage later in the England of Queen Victoria and is now enjoying a strong resurgence.

You can play it very comfortably on a chess board with 32 pennies. You can use different coins, since the color and size make no difference. Use pieces of paper to cover any unnecessary squares to come up with the following figure:

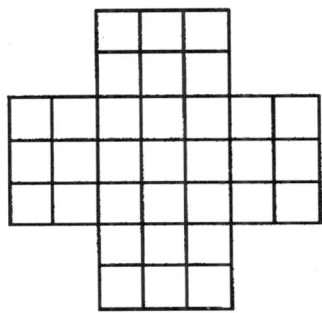

Now place the 32 coins, one in each square except for the one shown here, like this:

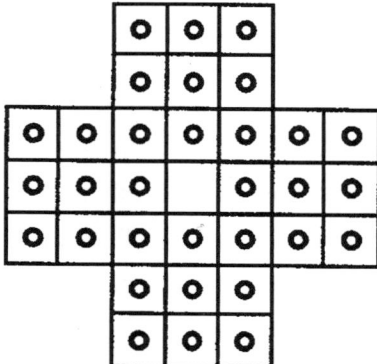

Now you may begin moving the coins. One coin may jump over another next to it by moving left or right, up or down, as long as the "jumped" coin has an empty space next to it. So the "jumped" coin gets "eaten," or is removed from the board. For example, you may go from this situation

to this one

or from this

to this

The idea is to try to end up with the board looking almost the exact opposite as when you started, or like this:

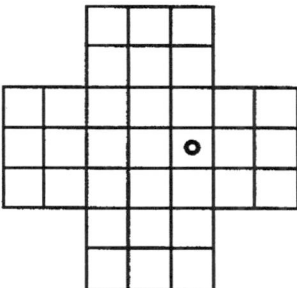

This game is fun and not at all easy. Get to work on it, moving your coins from one square to another. Now stop reading, close your book, and come back only after you've spent a good while having fun...

Now I hope you'll forgive me for having played a little joke on you. The game as proposed here is impossible. How do I know? Maybe you've found another way of looking at it in the time you were playing (if so, congratulations!; please write and tell me about it), but this is the method that Allen Shields from the University of Michigan told me, using the **Klein group**. What is it? It's the commutative group of four elements we'll call a, b, ab, 1 (we could call the third element (ab) c, for example, but by calling it ab the multiplication table for this group will be easier to memorize). The table for the group is as follows:

•	a	b	ab	1
a	1	ab	b	a
b	ab	1	a	b
ab	b	a	1	ab
1	a	b	ab	1

(As you can see, we only have to remember that $aa = 1$, $bb = 1$ and that 1, naturally, is the group unit.)

What does this have to do with solitaire? Actually, it is related to a number of games and types of solitaire, as you'll see later. We can use our board to write in the elements of the Klein group as follows:

			ab	a	b		
			b	ab	a		
b	ab	a	b	ab	a	b	
a	b	ab	a	b	ab	a	
ab	a	b	ab	a	b	ab	
			a	b	ab		
			ab	a	b		

Note these two things:
(1) *The product of all the elements of the group corresponding to the squares where the pennies are located at the beginning of the game is* a. This is obvious because every three contiguous squares contain a, b, ab and $abab = 1$. This way it's easy to find the product indicated, which is a.
(2) Any movement of pennies allowed in our solitaire may be interpreted as follows. Let's take a look at this move, for instance:

According to the way we've named the squares with the elements of the Klein group, note that the element of the group in the square that's occupied after making the move is the product of the two elements of the group in the squares that were occupied before moving. So we see that *the product of the elements of the group pertaining to squares occupied before making a move is the same as the product of the elements pertaining to the squares occupied afterwards, meaning that this product does not change over the course of our moves.*

With these two observations it is clear that this solitaire game is impossible. At the beginning, the product of the elements of the group pertaining to occupied squares was a, and, in the game proposed here, it should be b, but that is impossible.

The solitaire game of the gentleman from the Bastille wasn't quite so unfair. The final position is supposed to be this one, where, as you can see, the product that corresponds to the only element that should be left is a, like at the beginning.

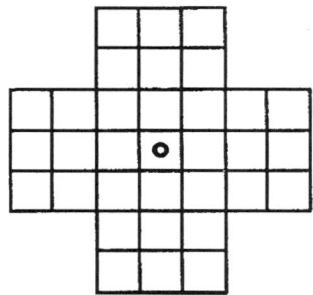

Now things are really getting serious. Get ready for a good time. Close the book and get going. You might just work it out. The hard part, if you do manage, is to remember how you did it. If you don't manage and want a hint or you want to impress your friends with your clear-sightedness, come back to the book.

Here's a strategy made up of a bunch of *macro-moves* that allow you to take giant steps. Each macro-move is a sequence of obvious moves in which a basic set of squares with gamepieces in a given catalytic situation next to it can be emptied without changing the catalysts.

Macro-move a. From a partial set-up like this one:

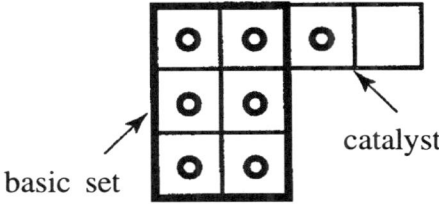

you go to the following set-up:

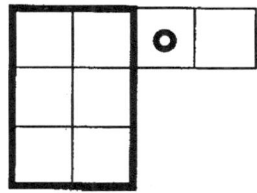

Macro-move b. From this set-up:

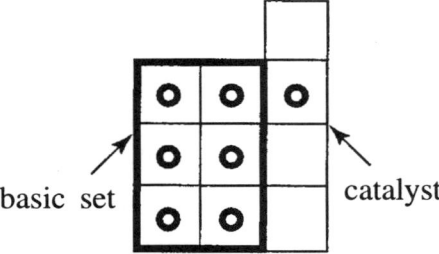

you go to this situation:

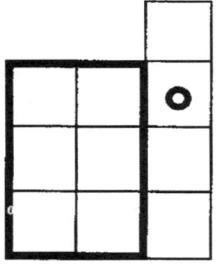

Macro-move c. From this set-up:

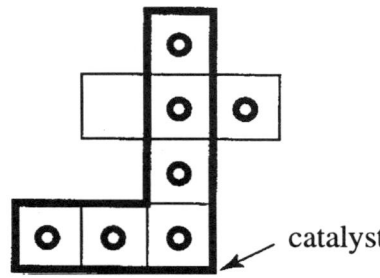

you go to this one:

"Solitaire" Confinement 49

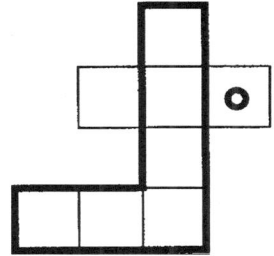

Macro-move d. From here:

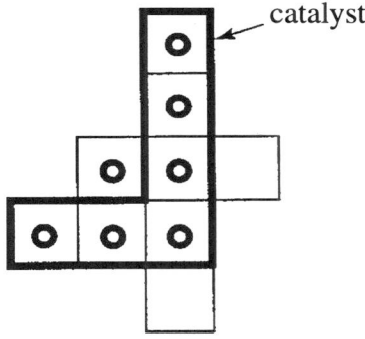

you go to:

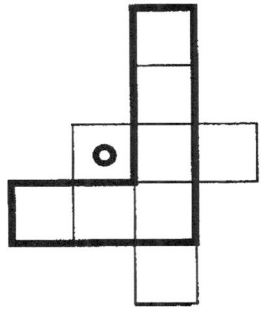

To use these macro-moves, let's look at the initial board after one move, like this:

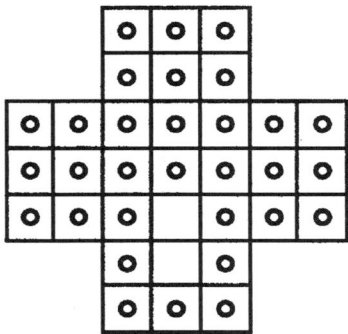

And now, I'll tell you in order the macro-moves you can use to take you to a happy ending.

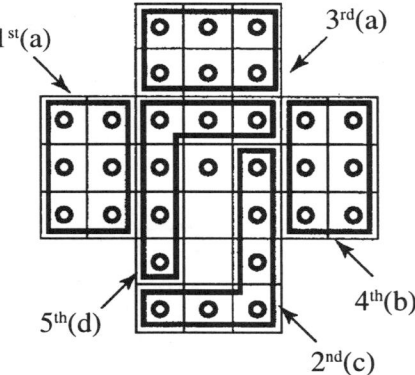

One question, out of curiosity. Using the same set-up, you're asked to end up like this:

"Solitaire" Confinement 51

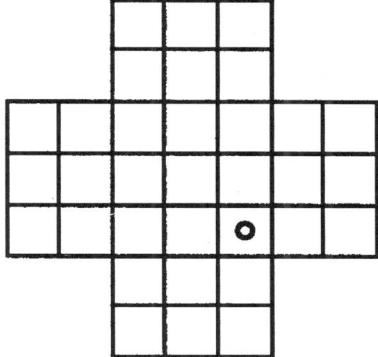

or like this:

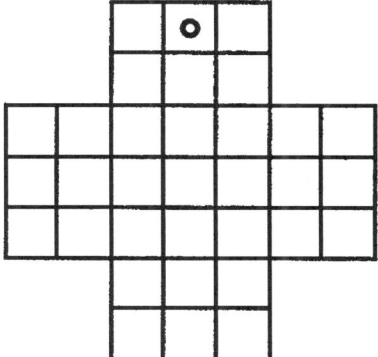

As you can easily see, the obstacle that came up at the beginning when I tricked you doesn't seem to exist here. The product at the beginning was *a* and it is still *a* at the end. Can this solitaire puzzle be solved under these conditions? Can you find the recipe, if it exists? If you find it, please send it to me.

Notes

"I enjoy the above solitaire game very much. I play it in reverse order. That is, instead of trying to create a certain figure (such as leaving one man in the center square) according to the rules of the game, that is, jumping over a man to an empty spot and removing the jumped man, I consider it better to play it backwards: starting from a given figure, jumping over an empty space and putting a man in the empty space just jumped."

These are the words of the great Leibniz, in a letter from 1716. Leibniz was convinced of the tremendous educational value of games, as he explicitly expressed in another letter of 1715:

"Men are never more ingenious than in the invention of games. There the spirit feels quite at home... After the games that rely solely on numbers come those in which the situation plays a part... After the games in which only the number and the situation play a part would come the games in which movement plays a part... In sum, it would be desirable for an entire course to be done on games treated mathematically."

Gottfried Wilhelm Leibniz (1646–1716) was one of the most profound and multifaceted spirits in the history of mankind. A deep philosopher, logician, mathematician, scientist, expert on law, history, linguistics and theology, he was also a tireless worker in the religious and political reconciliation in the Europe of his time, which was divided in so many ways.

He was born in Leipzig, where he initially trained in the library of his father and later at the university of his home town. He first studied law and then became especially interested in mathematics and metaphysics, one of his spiritual ideas being to make a methodological synthesis that would allow the entire field of knowledge to be addressed using mathematical methods. He placed himself at the service of the prince elector of Mainz, thus starting off on his career as a politician, which he juggled with a multitude of research of the most diverse nature. Few scientists have managed to combine such an active life with the same depth of thought revealed in the philosophical, mathematical, logical, historical and other writings of Leibniz. At the service of the Duke of Hannover, Leibniz tirelessly traveled throughout the countries of Europe in his stagecoach, which was at the same time the place he used for thought and work.

The Mathematician as a Naturalist

How do you think that theorems are invented, those horrific theorems that are sometimes given to you out of nowhere just like a meteor shower? Even though most of the letters are the same, it turns out they are exact opposites, and in this little essay, one thing I would like to make perfectly clear is precisely this theorem:

THEOREMITE ≠ METEORITE

Pythagoras used to love playing with little stones (calculus) and making calculations with calculi. One of his games was the following:

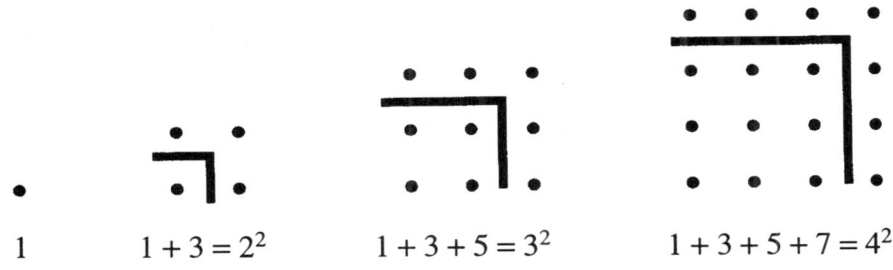

$$1 \qquad 1+3=2^2 \qquad 1+3+5=3^2 \qquad 1+3+5+7=4^2$$

"Aha!" he said to himself, "I've got the trick! The sum of the first n odd numbers is n^2."

This is a theorem that emerged from the simple stones from a beach... and the observations of Pythagoras. Is this a proof? If it isn't, it must be close, right?

Euler looked and looked again with great attention and pleasure at his collection of Platonic solids, regular polyhedrons, like someone gazing at a collection of precious stones. There he had the pointed tetrahedron, the cube with its solid appearance, the octahedron, the strange dodecahedron and the mess of the icosahedron, all the same on all their sides. All full of symmetry and regularities. It occurred to him to count their faces, their vertices, their edges. Maybe there was a law linking these numbers together! In such a perfect world these numbers should also be governed by a simple, transparent law!

Why don't you make your own collection of perfect solids? It's easy. For example, cut a very regular pentagon out of a piece of construction paper. The easiest way to draw a regular pentagon is to make a knot with a strip of paper like I've shown in the figure below. Why don't you try to prove that if you start with a perfectly rectangular strip, what you end up with is a regular pentagon? It's fairly easy to do.

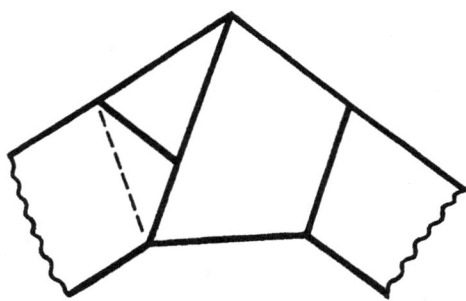

With the help of this pentagon cut-out, draw twelve pentagons like this one together on a piece of construction paper, as shown below (I've drawn smaller ones so they'll fit on the page).

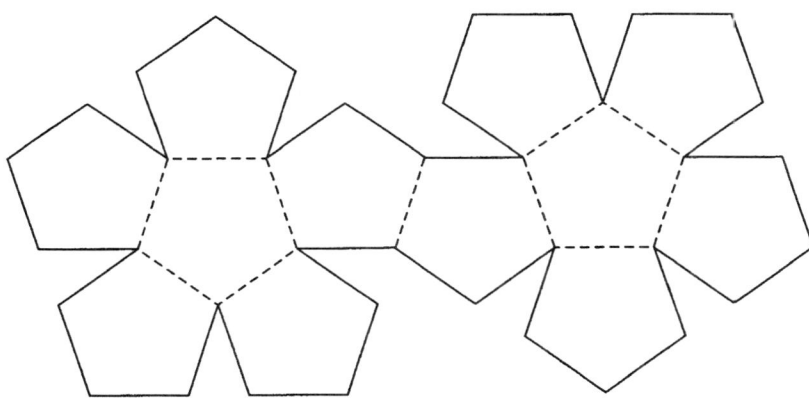

Now cut along the outer lines, leaving little flaps so you can glue the faces together on the edges, like so:

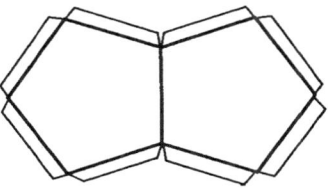

Now fold along the edges of the pentagons (if you want to be completely accurate, it's best to score the edges with a knife) and try to paste them together to form a closed polyhedron. If you've drawn them accurately, you'll see (wonder of wonders!) that they fit together perfectly and the flaps can be glued to make a figure like this one:

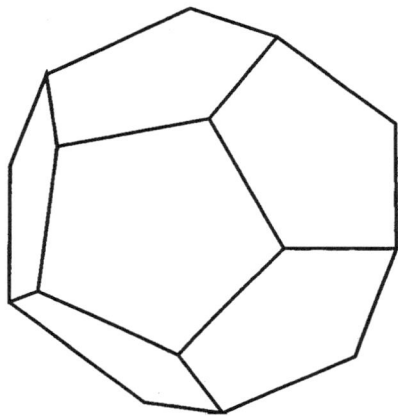

Don't you think it's wonderful that they fit together so well? Do a little experiment. Instead of starting from a regular pentagon, do the same thing with a regular hexagon. Draw a bunch of regular hexagons bunched up together and…, watch what happens. Starting with a regular heptagon things get even worse. But starting from regular polygons with fewer sides you can construct closed figures. Experiment a bit — experimentation is the mother of science. Starting from an equilateral triangle, using four, you can make a tetrahedron, with eight a regular octahedron, and with twenty an icosahedron. Starting with a square, six equal ones will make a cube, and from a regular pentagon, twelve will make a regular dodecahedron. What's the mystery? Are there more regular polyhedrons? Try to study this from a geometric point of view, which is how the ancient Greeks did it. Why is it that with regular equilateral triangles, you can make three different regular polyhedrons, but with squares or pentagons you can make just one?

Count, like Euler did, and make up a list of the vertices, edges and faces of your regular polyhedrons.

	Faces	Vertices	Edges
Tetrahedron	4	4	6
Cube	6	8	12
Octahedron	8	6	12
Dodecahedron	12	20	30
Icosahedron	20	12	30

"What strange numbers!" Euler said to himself when he had finished. But numbers didn't scare Euler. He began to do calculations with them. "There are always more edges than vertices and faces... naturally! If we begin with an S-agon (a regular polygon with S sides), in order to count the edges E, we can multiply the number of faces F by S. But then we would count each edge twice. Clearly, $FS = 2E$. Since S is at least 3, it must be true that $E > F$. Let's see if the $FS = 2E$ formula works...

| Tetrahedron | $FS = 4 \times 3$ | $2E = 2 \times 6$ |
| Cube | $FS = 6 \times 4$ | $2E = 2 \times 12$ |

it's all working out like I thought."

"Could a similar relationship be established between edges and vertices? It seems like something like that could be done. I'll count the number of edges meeting at a given vertex G. So to count the edges we can multiply V, the number of vertices, by G. But that means we've counted the number of edges twice. So $VG = 2E$. Let's see:

Tetrahedron	$VG = 4 \times 3$	$2E = 2 \times 6$
Cube	$VG = 8 \times 3$	$2E = 2 \times 12$
Octahedron	$VG = 6 \times 4$	$2E = 2 \times 12$

Everything is going as planned. Since G is always at least 3, then $E > V$."

The list of numbers from page 59 is interesting, isn't it? There are strange relationships. Look:

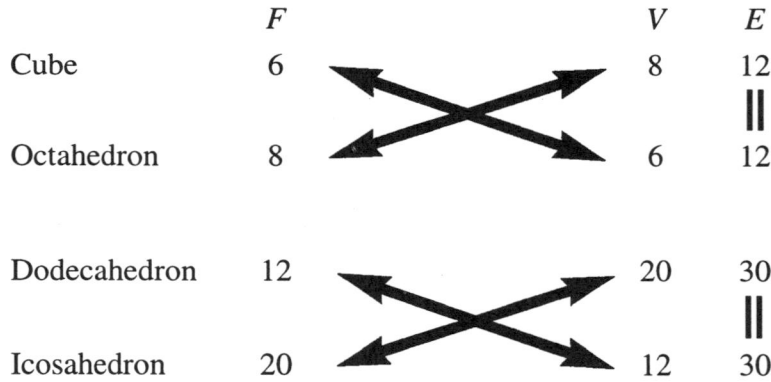

And, if we look at the values for S and G, it turns out that:

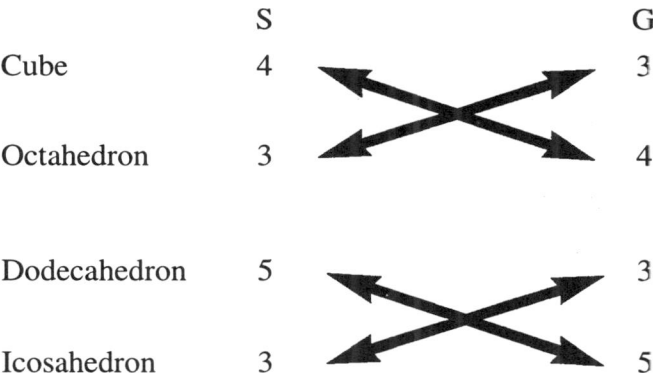

These relationships would seem to suggest that the way the cube and the octahedron behave towards each other is similar to the way the dodecahedron behaves towards the icosahedron. Faces are exchanged for vertices and vertices for faces. What explanation can be given for this strange phenomenon?

On each face of a regular polyhedron there is a special spot, the center. Could it be that the centers of the faces of an octahedron...? Yes!! EUREKA!! They are the vertices of a cube! And the centers of the faces of a dodecahedron are the vertices of a regular icosahedron! And vice versa! We've just solved an enigma. This relationship fully explains the above diagrams. It even explains why the number of edges is the same, 12, in cubes and octahedrons, and 30 in icosahedrons and dodecahedrons. The *FS* of one is the *VG* of the other, and they're both 2*E*.

The introduction of the numbers *G* and *S* has complicated our initial *F*, *V* and *E* table and clarified a few things. But can we find a direct relationship between *F*, *V* and *E*?

By playing a little with the numbers, the chart on page 59 suggests the following relationship:

$$F + V = E + 2$$

which checks out in all cases. Coincidence? How far can we go with this relationship? Let's construct another non-regular polyhedron, a prism or a pyramid, for instance.

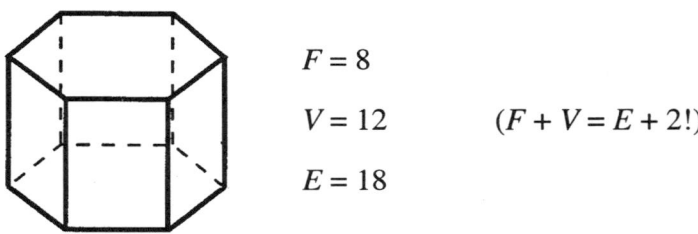

$F = 8$

$V = 12$ $\quad (F + V = E + 2!)$

$E = 18$

It would seem that for any polyhedron of this kind we find that $F + V = E + 2$. This was more or less the line of thought of the great Euler, and the following is one of the many famous theorems that bear his name:

FOR ANY CONVEX POLYHEDRON

$$F + V = E + 2$$

Euler never proved his theorem. Nearly a century went by before Cauchy hit upon a proof that works using the following ingenious idea. Suppose we have a polyhedron wrapped in a very stretchy mesh that clings perfectly to its faces. Let's draw the edges and vertices on it very clearly, cut out one of the faces with the corresponding bit of mesh, remove it and, making use of its "stretchiness", spread it out on a plane so that the edges don't get mixed up. The edges don't have to be straight. For example, from a cube we get:

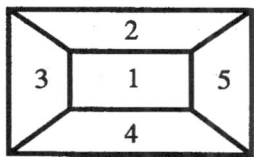

from an octahedron we get:

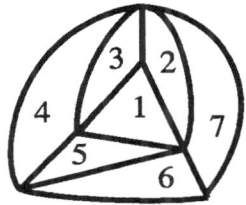

from a dodecahedron we get:

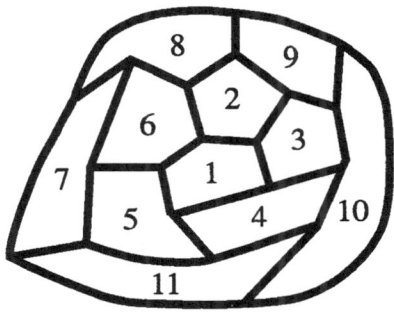

If we prove that, for any figure of this type it can be verified that $F + V = E + 1$, then we'll have a demonstration of Euler's theorem.

What is a figure of this type? It is clearly a figure made up of a set of polygons with curved lines such that it leaves no holes inside — in other words, *it is a curved polygonal configuration the inside of which has been divided up by non-overlapping curved polygons.*

Now we can begin to make an induction on the number of polygons that make up the division of the configuration, based on the observation that if we eliminate a polygon from the outside edge of figure of the type described, what we are left with is a figure of the same type, only with one polygon less. If the number of polygons is 1, then $F = 1$ and, since $E = V$, we obtain that $F + V = 1 + V = 1 + E$.

Now let us suppose that in the case of any configuration of h or fewer polygons, we know that $F + V = E + 1$ (inductive hypothesis). Let's consider any case of a configuration where the number of polygons F^* is $h + 1$. Let E^* and V^* be the number of edges and vertices. If from this configuration we remove one of the outside faces, we get a configuration of the type we are studying with h polygons. For this configuration, the inductive hypothesis gives us $F + V = E + 1$. Now we have $F^* = F + 1$, $V^* - V = E^* - E - 1$, and thus $F^* + V^* - E^* = F + V - E = 1$, that is to say, $F^* + V^* = E^* + 1$. With this we have proven the theorem. (Careful! It could be that by removing an outside face you end up with not one, but two separate configurations of the type we are studying. How do you think the proof would have to be modified?)

The theorem is pretty ingenious, isn't it? Let's also take a look at how powerful it is, in order to solve our still unsolved mystery:

Do more regular polyhedrons exist?

We know that $F + V = E + 2$, $FS = VG = 2E$.
So, $F + FS/G = FS/2 + 2$.
In other words, $1 = S(1/2 - 1/G) + 2/F$

Now, we know that $G \geq 3$, $S \geq 3$. Let's try with different values.

$G = 3 \to 1 = S/6 + 2/F$
- If $S = 3$, $F = 4$ Tetrahedron
- If $S = 4$, $F = 6$ Cube
- If $S = 5$, $F = 12$ Dodecahedron
- If $S = 6$, $F = -2$ impossible
- If $S > 6$, $F < 0$ impossible

$G = 4 \to 1 = S/4 + 2/F$
- If $S = 3$, $F = 8$ Octahedron
- If $S = 4$, $F = -2$ impossible
- If $S > 4$, $F < 0$ impossible

$G = 5 \to 1 = S/5 + 2/F$
- If $S = 3$, $F = 20$ Icosahedron
- If $S = 4$, $F = 5/2$ impossible
- If $S = 5$, $F = -2$ impossible
- If $S > 5$, $F < 0$ impossible

$G = 6 \to 1 = S/3 + 2/F$
- If $S = 3$, $F = -2$ impossible
- If $S > 3$, $F < 0$ impossible

$G > 6$, $S \geq 3 \to F < 0$ impossible

Therefore, the only regular polyhedrons that exist are *the tetrahedron, the cube, the octahedron, the dodecahedron and the icosahedron.* There's no need to look any further; they can't exist. As you can see, this method of demonstration proves more than the geometric method. If you tried to establish this impossibility using the geometric method, you'll understand why this is so. This method proves that even if you remove the condition of regularity and require only that the faces have the same number of sides, even if they're not regular, we find that there cannot exist polyhedrons that satisfy this condition other than tetrahedrons, hexahedrons, octahedrons, dodecahedrons and icosahedrons, which are deformations of the regular polyhedrons. You see, then, that the method we have used here did not take equal size of faces, sides and angles into account at all. Later on we'll see other applications of Euler's theorem.

It's really odd to realize that in three-dimensional space there are only five regular polyhedrons known by Pythagoreans. In four-dimensional space there are six regular polyhedrons (here they are usually known as regular polytopes). In contrast, in five-dimensional space there are only three (corresponding to the tetrahedron, the cube and the octahedron).

Notes

One of the mathematicians who best combined the audacity of mathematical thought with respect for observed phenomena was Johannes Kepler, the discoverer of the fundamental laws of the movement of the planets around the Sun.

Kepler was born in 1571 in Wil der Stadt, one of the "free cities" of the Holy Roman Empire of the German Nation. Born prematurely into a poor family, Johannes was small and rather sickly, but his extraordinary intelligence manifested itself early on. Thanks to the Duke and Duchess of Württemberg, he began his university education in 1587 at the University of Tübingen, where Michael Mästlin, professor of astronomy, introduced him to the Copernican ideas on the solar system.

In 1591, after finishing his degree, he began to study theology in Tübingen with the idea of becoming a Lutheran minister, but in 1594, due to the death of a mathematics professor in the Austrian city of Graz, Kepler was appointed to fill the vacancy.

The most striking characteristic of Kepler's way of thinking was the deep conviction, already present in the ancient world in Pythagoras and Plato, that the entire universe is decipherable through mathematical thought. There were six known planets in his time. There were five possible regular solids, as we have seen. By making them the right size, it's possible to place all of them concentrically in such a way to allow six spheres to alternatively inscribe and circumscribe the regular solids. This was the deep reason for which there were six planets and not fifteen or twenty-five! The existing planets revolved around the Sun in these six spheres! Besides, this allowed the distance from the planets to the Sun to be estimated. It is interesting that, although this initial hypothesis was wrong (there are a few more planets than Kepler came to know), the distances calculated did not turn out to be as absurd as one might have first thought. Later, Kepler would abandon his hypothesis due to the impossibility of using circles to figure out the orbit of Mars, which was one of the very particular objects of both his observations and those of Tycho Brahe, the imperial astronomer who took him on as part of his team. But the conviction that mathematics were involved in all shapes led him to think of an elliptical orbit and to formulate his three laws: a) planets follow ellipses, one of whose foci is the Sun; b) areas swept by the segment that goes from the Sun to a single planet in the same time are equal; c) the cube of the major semi-axis of the ellipse that a given planet follows divided by the square of the time that the planet spends in following it is a number independent of the planet chosen.

Kepler's deep faith in the mathematical intelligibility of the universe produced the miracle of the founding of these three laws! And they served Newton later to convince the world of the validity of his universal law of gravitation when he mathematically deduced Kepler's three laws based on his law of gravity.

Four Colors Are Enough

Some people think computers are idiots. You have to tell the whole joke to them and explain it all to them before they'll laugh. A comma instead of a semi-colon and they've had it. The truth is, pens are even dumber, but without their help there wouldn't be many problems we could solve.

The story I'm going to tell you is about a problem that was solved only recently, and with the indispensable help of a computer. And it wasn't used just to count fast, either. There are probably a lot of mathematical results that lie deep in slumber in our computer circuits just waiting for the magic program to wake them up.

A Bit of History

In 1852, Francis Guthrie, who had recently graduated from the University of London, wrote to his brother, still a student there, asking if there might exist a proof of the fact used by map printers that four colors were enough to color any given map. Frederick, his brother, was unable to give him an answer, but asked one of his professors, a good mathematician and also fond of puzzles and mathematical games. His name was Augustus De Morgan. De Morgan was unable to prove it, but passed the buck, which finally reached one of the most famous mathematicians at the time, Arthur Cayley. In 1878 Cayley proposed the problem to the London Mathematical Society as an interesting one. Barely a year later, a London lawyer, Arthur B. Kempe, published an article putting forward a demonstration that four colors would suffice. Kempe's solution was accepted as good for eleven years. In 1890, P.J. Heawood found a flaw in Kempe's ingenious and complicated argument. Heawood got so excited about the problem that he dedicated his entire life to studying it in depth. For more than sixty years he worked on it from a wide variety of angles, obtaining interesting results that helped topology move forward considerably, but he never managed to solve the original problem. Among other things, he found out that for any given map on the surface of a tire, seven colors are enough, that for a map on a Möbius strip, six would suffice... problems that were apparently more complicated were solved quickly, but Francis Guthrie's problem on the globe would have to wait more than one hundred years for the decisive help of the computer!

In 1950, it was known that any map of less than 36 countries could be colored with four colors. In the fifties, Heinrich Heesch, a professor in Hannover, began to think that Kempe's ideas, together with the aid of the computer, could perhaps lead to a solution, but although he had an idea about how it could be done, he was still a long way from realizing his plan. From 1950 to 1970, Heesch developed the techniques that led to the solution. This work plan was perfected and carried out from 1970 to 1976 by Kenneth Appel and Wolfgang Haken of the University of Illinois. After many hours of thought and dialog and work with the computer, they were finally able to announce, in June 1976, that four colors were indeed enough.

I'll try to give you an idea of the points that have marked the way towards the strange manner of solution that we now have.

The Problem

The idea is to find the minimum number of colors necessary to properly color in any type of map on the globe or in the plane. (This number will be the same for both cases, globe and plane.) Countries with a common border should be given different colors. The case of a country with separated bits within other countries (enclaves) is excluded. It is easy to see that, by setting a number such as 6, for example, it is possible to create a map requiring six colors. So, we are excluding the following type of situation:

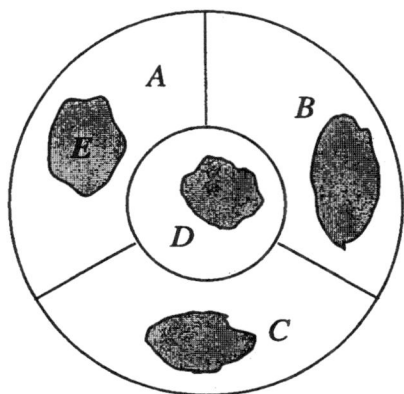

For this map, where E is the country shaded in gray with enclaves in all the others, you need five colors.

It's also obvious that three colors are not enough to color any map. This one, for instance:

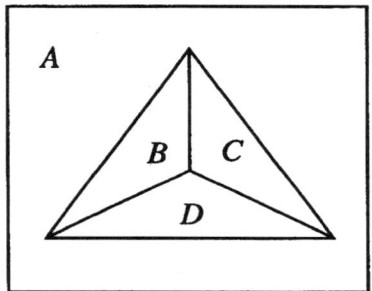

requires four.

Heawood proved very quickly that 5 colors were enough. But would 4 suffice?

For some of the reasoning we'll be using, it's a good idea to give the problem a more convenient shape that's equivalent to the one we have seen. For any given map we can make a corresponding graph as follows. (Remember that a graph is simply a set of points, or vertices, some of which are joined by lines, or arcs.) We do it the following way. On the inside of each country we mark a capital. The capitals will be the vertices of the corresponding graph, a dual graph. If two countries have a common border, we connect their capital by a highway that crosses the common borderline without the highways' crossing each other. These highways are the arcs of the dual graph.

For example, if our map is like this

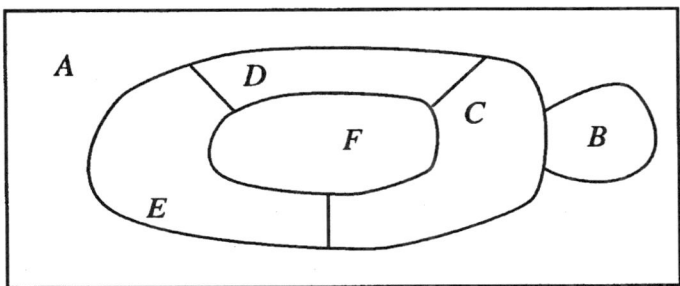

we mark the capitals and connect them according to this rule:

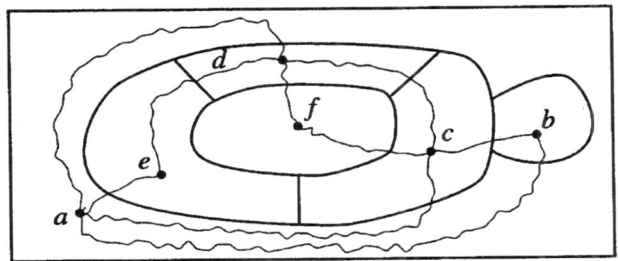

This gives us the following dual graph:

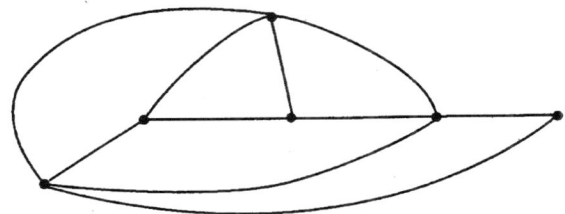

Countries correspond to capitals and to vertices of the graph
Borders correspond to arcs of the graph

The valence of a vertex (capital) of the dual graph, that is, the number of arcs that meet there, is the number of countries that correspond to that capital (vertex).

The problem, in terms of the dual graph, consists of determining the minimum number of colors to shade a graph similar to that resulting from a map such as the one described so that two adjacent vertices have different colors.

The proof that Kempe gave in 1879 that four colors were enough was certainly ingenious and, although wrong on one point, it served as the basic outline for the proof we now have. That's why it's worth showing it here, since it will then be easy to understand today's approach towards the theorem.

Kempe's "Proof"

We'll call any map that *requires five colors* to be properly shaded a "penta map" (pentachromatic map, if you don't want to shorten it). This map cannot be colored with less.

Our objective is to prove that no penta map exists.

We'll call any map that fulfills the following conditions a "normal map": a) it contains no isolated country inside of another, i.e., no country with only one neighbor; b) there is no border point that borders more than three neighboring countries.

Situations like the ones below are, therefore, excluded:

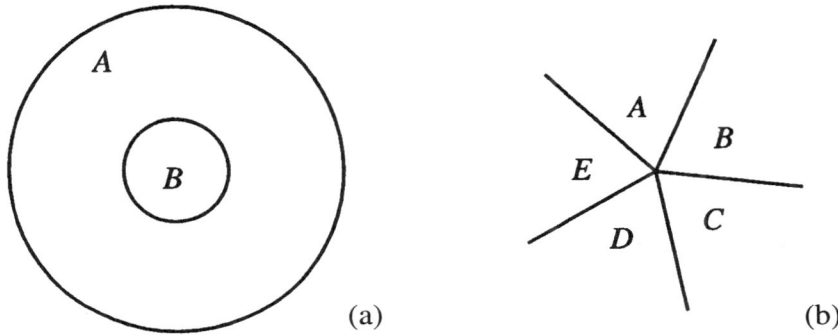

(a) (b)

Note that the graph corresponding to a normal map (the border points are borders for two or three countries at the most) is a curvilinear triangulation of the globe, that is, a partition of the globe into curvilinear triangles.

Now, the four following steps will allow us to understand Kempe's proof that no penta map exists:

1) *If a penta map M exists, then a normal penta map exists.*

So if *M* has partial configurations like this one:

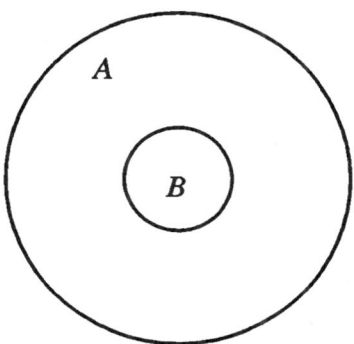

they may be substituted by others like this one:

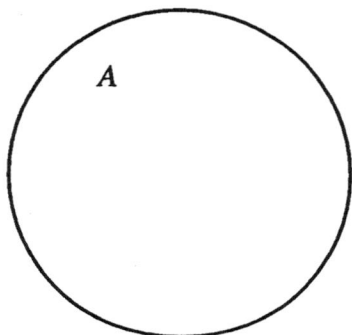

and if it has a partial configuration of this type (a border point with more than three countries),

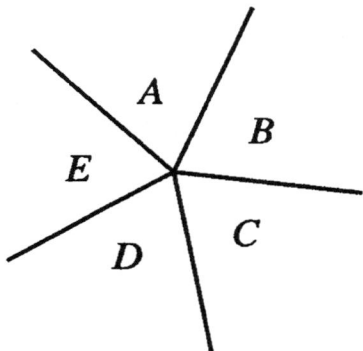

it is substituted by another one like this (by adding a country we find that no border point borders on more than three neighboring countries).

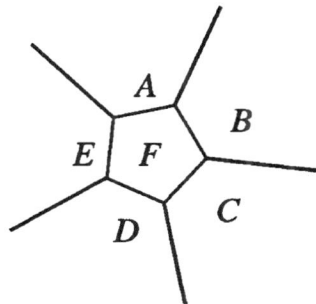

Let's call the new map, which is clearly normal, M^*. If the new map M^* turns out not to be penta, we could color it with four colors. Now we color it. But then, by adding the countries we've deleted from M to obtain M^* and removing the ones we've added, it is easy to find that M may also be colored with four colors, unlike what we had supposed.

2) *If a normal penta map exists, there exists a normal, minimum penta map*, i.e., with the lowest possible number of countries. Let's consider all normal penta maps. Each has a finite number of countries. One of them must have the lowest number. It's obvious, right?

3) *Any normal map contains at least one country with less than six neighboring countries.*

Now we're getting into more serious stuff and the problem is getting deeper. This step by Kempe was also seen to be completely valid.

To better understand this step, let's take the normal map, which we know is a partition of the sphere into curvilinear polygons. To this we can apply Euler's theorem, which you already know:

Faces + Vertices = Edges + 2

We can express the number of faces, F, or countries, as follows. If F_2 is the number of countries with two neighbors, F_3 the number of countries with three neighbors, etc., then clearly $F = F_2 + F_3 + F_4 + \cdots$

Furthermore, each arc or edge is the border of two neighboring countries, so

$$2F_2 + 3F_3 + 4F_4 + \cdots$$

is the number of arcs counted twice, or

$$2E = 2F_2 + 3F_3 + 4F_4 + \cdots$$

Also, the map is normal and so exactly three border arcs meet at each vertex. That is why $3V$ is also the number of edges counted twice. In other words, $3V = 2E$. By eliminating F, E and V in

$$F = F_2 + F_3 + F_4 + \cdots$$

$$2E = 2F_2 + 3F_3 + 4F_4 + \cdots$$

$$2E = 3V$$

$$F + V = E + 2$$

we easily obtain that

$$12 = (6-2)F_2 + (6-3)F_3 + (6-4)F_4 +$$
$$+ (6-5)F_5 + (6-6)F_6 + (6-7)F_7 + \cdots$$

Therefore, since the sum is 12, at least one of the numbers F_2, F_3, F_4, F_5 is greater than zero, which expresses precisely what we are trying to demonstrate. Pretty tricky, huh? This relationship by Kempe proved to be very useful in the proof achieved in 1976.

4) *No normal, minimum penta map may contain a country with less than six neighboring countries.*

If we could demonstrate this, we would have a proof that penta maps cannot exist, since this contradicts step 3, which we have already established. Kempe made a mistake on this point, but at the very end. A good part of his reasoning is also valid and served in the correct proof.

To prove step 4 we begin by taking a normal, minimum penta map M. If it were to contain this partial configuration:

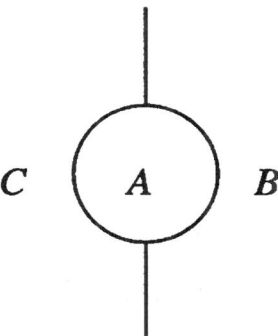

country A with two neighbors, the map M^* that would result from substituting the above partial configuration with the following one (deletion of A) in M

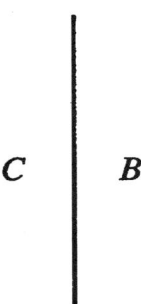

is normal and is not penta, since it has one country less than M, and M was a normal, minimum penta map. So M^* can be shaded using four colors. Now we color it. But now it is clear we can also color M with four colors, returning A to its original place and giving it a color other than the ones given to B and C. This contradiction proves that M cannot have the assumed configuration, if it exists.

The following configuration (A with three neighbors)

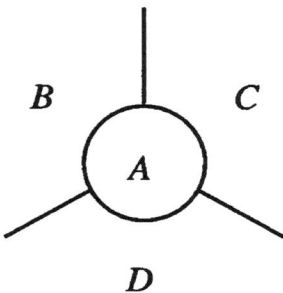

is excluded in the same way, deleting A, coloring in the resulting map, M^*, and then coloring M.

Excluding this configuration is more complicated:

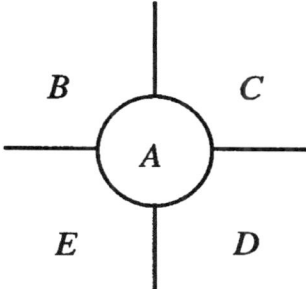

Let's go to the dual graph to reason it out. The dual graph is a triangulation of the sphere that contains this partial configuration:

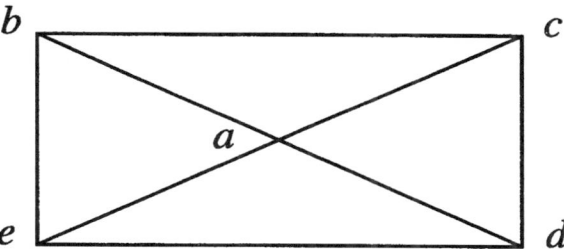

We eliminate country *a* by making *a* identical to *b*, and thus the *reduced* configuration is as follows:

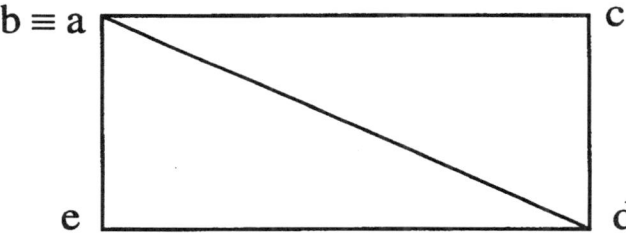

The resulting graph is not penta. It can be colored in with four colors, 1, 2, 3 and 4. We distinguish between two cases.

Case 1. If c and e have the same color, color 2 for instance, then b has another one, 3 for example, and d still another, let's say 4. So:

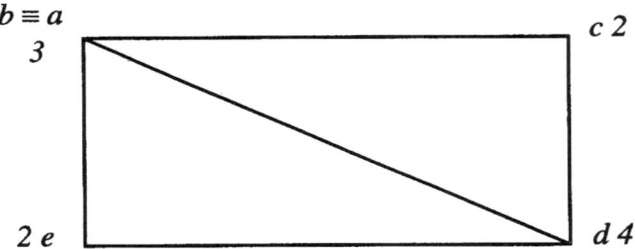

Then we reinstate a, give it color 1 and the initial graph is colored with four colors, contrary to the supposition that it was penta.

Case 2. If c and e have different colors, colors 2 and 3 for instance, then b and d have 1 and 4, as shown.

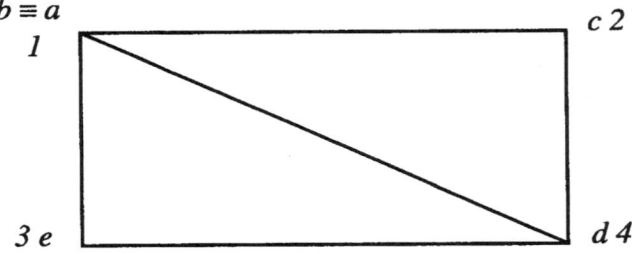

Then either we can go from c to e using arcs outside rectangle $bcde$ passing through a chain of vertices 232323...23 (Kempe's chain)

or we can't. If it is possible, then it is not possible to go outside of the rectangle from *b* to *d* through the chain 141414...14. We take the pair *bd* or *ce* for which it is not possible. Let it be *ce*, for example. Beginning at c there is a section 2323... that we cannot follow because we encounter the chain 141414...14 from *b* to *d*. We change this section to 3232..., which puts us into case 1, and we can proceed like we did there.

Kempe's flaw was in the reduction that he tried to make of the configuration below: in a similar manner, but more complicated. Here his proof was wrong. But from what we've seen the idea is clear. I'll write it again in modern terms, this time in three steps.

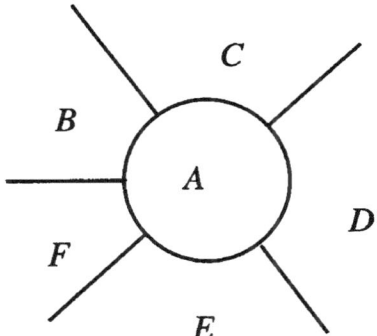

The Appel and Haken Proof

Having seen the outline of Kempe's proof, you'll now understand the following outline.

1) *Any normal penta map has certain inevitable sets of configurations.*

The inevitable set of partial configurations of Kempe was the following:

An inevitable set of configurations means that some of the configurations of this set have to be in the map.

2) *There are certain configurations that cannot be in a minimum penta map (reducible configurations).* Kempe proved that the partial configurations

are reducible. The proof he gave for

was wrong.

3) *The construction of an inevitable set of configurations, each of which is reducible, implies the non-existence of a penta map.*

In June 1976, Appel and Haken managed to construct a set of 1482 configurations, each of which was reducible.

Over the previous thirty years, enthusiasts of this problem, especially Heesch, had managed to come up with methods that could be programmed in a computer to construct inevitable sets of configurations. Programs were also made to check whether a configuration is reducible or not. These programs were long and complicated; so much so that even modern computers would need centuries to check whether the configurations of certain inevitable sets that had to be considered were reducible or not. Appel and Haken managed to reduce the task to more manageable dimensions. With this they finally obtained the result in 1976. The problem was solved.

The method consisted of an intelligent dialog with the computer. It is interesting to listen to them describe how, at a certain point in their work, the computer began to teach its teachers the right way to go about it.

"At this point the program, which had by now absorbed our ideas and improvements for two years, began to surprise us. At the beginning, we could check its arguments by hand, so we could always predict the course it would follow in any situation; but now it suddenly started to act like a chess-playing machine. It would work out compound strategies based on all the tricks it had been 'taught,' and often these approaches were far more clever than those we would have tried. Thus it began to teach us things about how to proceed that we never expected. In a sense, it had surpassed its creators in some aspects of the 'intellectual' as well as the mechanical parts of the task."

After hearing this you may ask yourself once again: Are computers smart or dumb? A well-taught computer can beat you at chess, beat you at Nim, beat you or at least tie with you at tic-tac-toe. And here we have the case of a theorem, the four-color theorem, which would never have been known as a theorem had it not been for the indispensable aid of a computer.

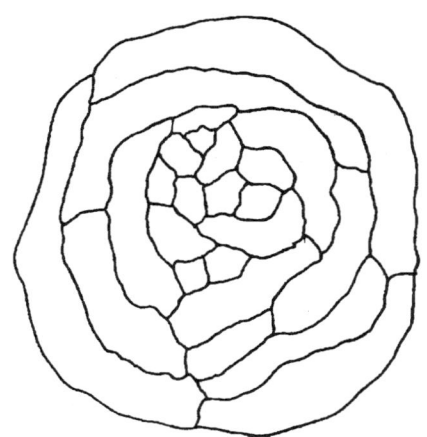

Could you color this map with four colors?

We'll probably be seeing lots of other theorems in the same vein soon. We'll also be seeing soon how a computer is capable of teaching you integrals, grading your homework and giving you a grade (ugh!). We hope that we can learn to use them well, because if they are used the wrong way they'll make our lives impossible.

Notes

The difficulty in coloring a given map properly shows up quickly in the following game invented by Stephen Barr. Two players, *A* and *B*, sit down with four different-colored pencils and a piece of paper. Player *A* draws a region. Player *B* gives it a color and draws a different region. Then *A* colors it and draws another region... The winner is the one who uses his wits in designing the successive regions to make the other one unable to properly color the region in question.

Leap Frog

You might be familiar with the game of leap frog.

If not, so much the better. To play you have to use a big piece of paper to draw a grid where each square is big enough to hold a penny. Draw a thick horizontal line five squares from the top of the grid, something like this:

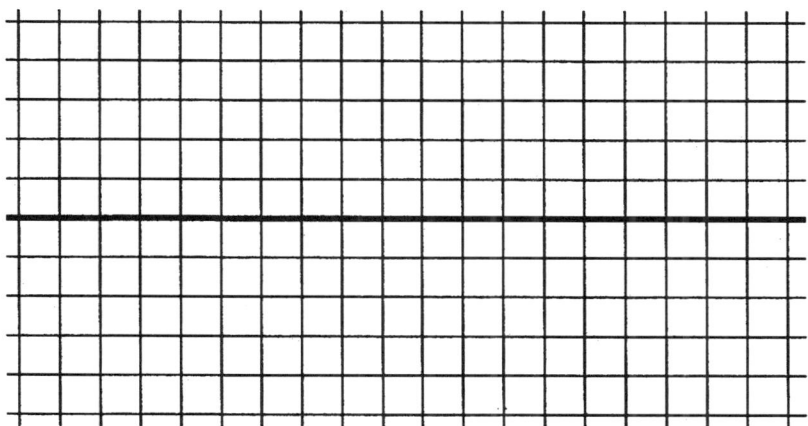

The game is played as follows. At the beginning you place a number of pennies (any number you want) distributed any way you prefer, each in one of the squares beneath the thick line. Once they're set up, you start moving and removing pennies from the board. Only horizontal (left and right) and vertical (up and down) moves are allowed, jumping over neighboring pennies as long as the square next to the "jumped" penny is empty, and removing the coin that was "jumped". For example, from this situation:

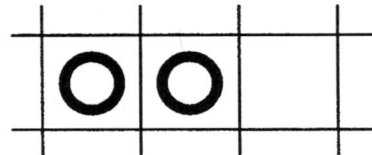

we could go to the following one:

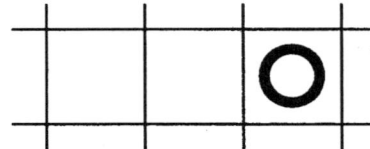

or from this one:

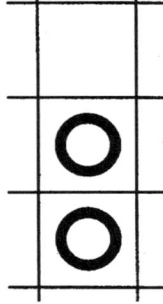

to the following:

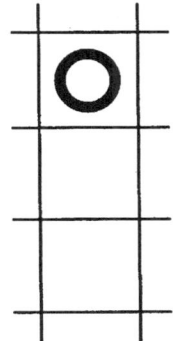

The idea is to set up the pieces or coins under the line in such a way that by using the allowed moves you manage to place a coin as high as possible above the thick line. When you've practiced a bit, you can try to do it with the minimum number possible of coins.

For example, to get to the first row above the thick line, it is clear that with just one coin it's impossible (there is no move allowed), but with two set up as follows:

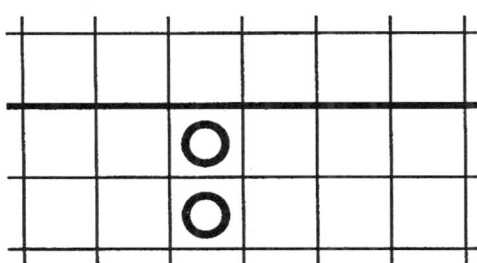

you can. The minimum sufficient number of coins to get one in the first row above the line is two. Get to work. How do you reach row 2? What is the minimum number? And what about to reach rows 3, 4, etc.? I don't want to rob you of the pleasure of finding it out for yourself. Close the book, and come back only after you've played for a while.

It probably didn't take long for you to discover that to get a penny into row 2, you have to get to this situation:

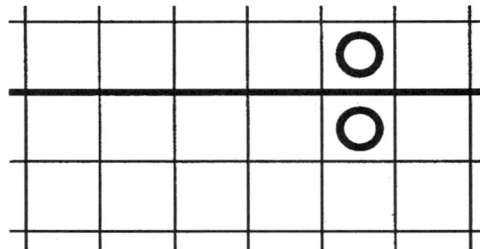

To get to it, all you have to do is to place four pieces in this position:

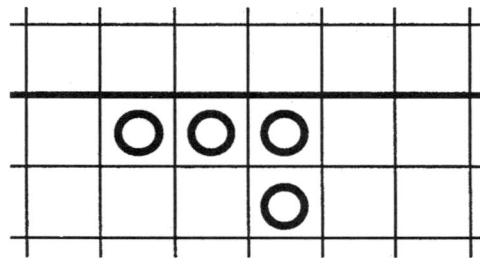

It is easy to see that you can't reach the second row with three coins. To reach row 2, the minimum sufficient number is 4.

And what about row 3? The idea would be to initially set up the pieces in order to get to this position:

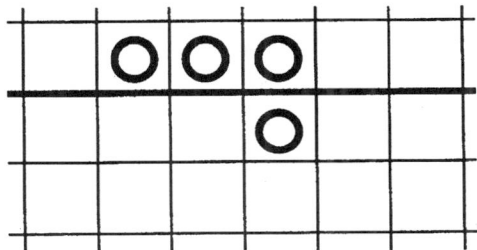

or a symmetric position, since we know that we can go up two more rows. How to do it? Since things are getting more and more complicated, the best thing to do is to try to invent some systematic means of proceeding. It might be reasonable to play backwards. In other words, our "frog" will now move two squares away down, right or left, leaving a penny in the middle square, always assuming it is empty. So, from this situation:

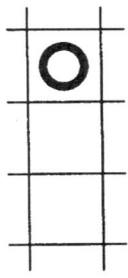

we could come to this one:

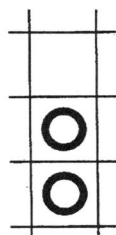

and from this one:

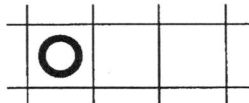

to this:

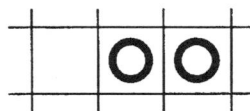

By means of these moves, the idea is to try to move the coins from this position

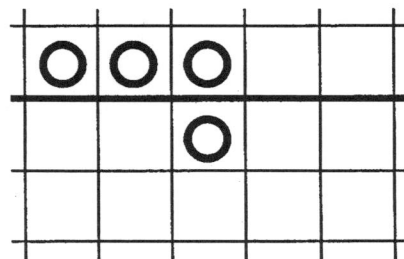

so that they are all below the thick line.

If you make a few attempts with this system you'll soon see that the situation you come to is this one:

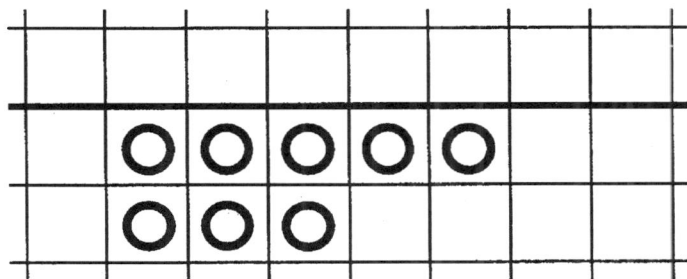

From this position, by playing forward, you can make it to the third row. There are eight coins. Can you prove that it can't be done with fewer?

It's interesting to observe that the third row can also be reached from the following situation, also using 8 coins,

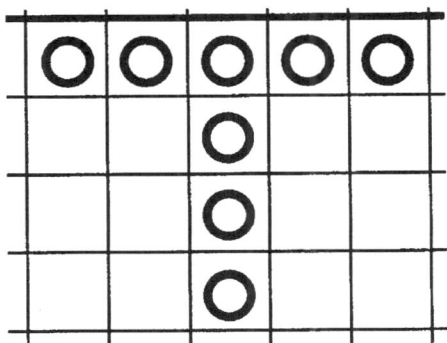

without the need for passing through the intermediate situation corresponding to the jump to the second upper row,

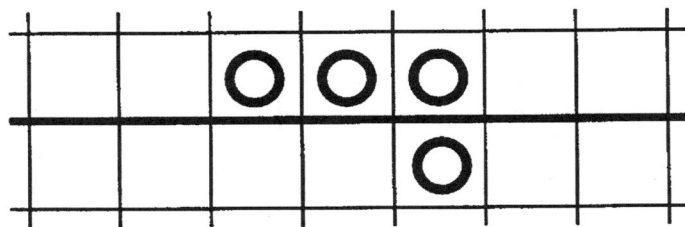

meaning that this initial 'T' position is not a result of the procedure of playing backwards.

Let's sum up our findings:

> For row 1, the minimum number is 2.
> For row 2, the minimum number is 2^2.
> For row 3, the minimum number is 2^3.

Now you're probably thinking that things would have to be pretty bad for row 4 not to have a minimum of 2^4, right?

Well, as bad as it may seem, that is exactly the case. To reach row 4, it is sufficient (not necessary, in principle, as we've seen happening with row 3) to start from an initial situation from which we can reach one of these two positions:

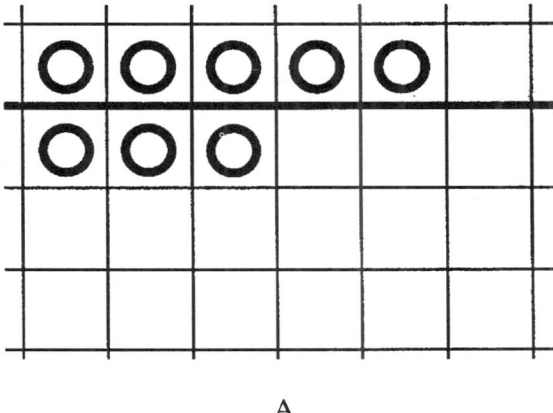

A

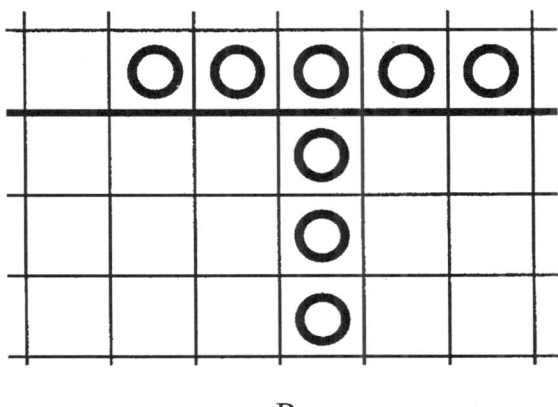

B

or, of course, a position symmetric to the first one.

Playing backwards and being careful not to add unnecessary pieces, you'll see that situation A and situation B can take you to the following initial situation:

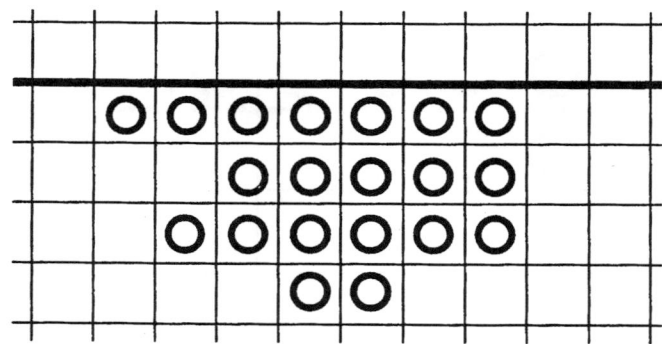

But, by playing backwards, situation A can also take you to this one,

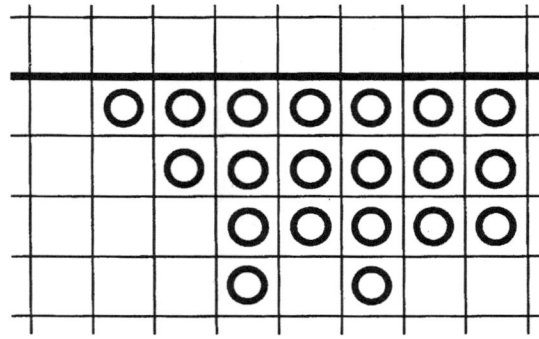

which is different from the previous one, but they both have 20 coins. Is it possible to get here with fewer coins? Why don't you try? And what about row 5? Go ahead. You'll see that with less than 20 coins you won't make it to row 4. For row 4 the minimum number is 20, which just goes to show you that you shouldn't rely too much on premature induction.

And what about row 5? We're no longer on solid ground for making a conjecture.

Well, though it may surprise you, row 5 cannot be reached with any number of playing pieces, now matter how you set them up.

The following proof is by John Conway, of Princeton University. We take the positive number w such that $w^2 + w = 1$. Mysterious, huh? Well, it turns out that, to make things even more confusing, w is the golden number $\varphi > 0$ such that

$$\frac{\varphi}{1} = \frac{1-\varphi}{\varphi}$$

In each box of our board, indefinitely extended downwards and to the right and left, we place a power of w as shown in the following figure:

—	—	—	w^4	w^3	w^2	w	1	w	w^2	w^3	w^4	—	—	—
—	—	—	w^5	w^4	w^3	w^2	w^1	w^2	w^3	w^4	w^5	—	—	—
—	—	—	w^6	w^5	w^4	w^3	w^2	w^3	w^4	w^5	w^6	—	—	—
—	—	—	w^7	w^6	w^5	w^4	w^3	w^4	w^5	w^6	w^7	—	—	—
—	—	—	w^8	w^7	w^6	w^5	w^4	w^5	w^6	w^7	w^8	—	—	—
—	—	—	w^9	w^8	w^7	w^6	w^5	w^6	w^7	w^8	w^9	—	—	—
—	—	—	w^{10}	w^9	w^8	w^7	w^6	w^7	w^8	w^9	w^{10}	—	—	—
—	—	—	w^{11}	w^{10}	w^9	w^8	w^7	w^8	w^9	w^{10}	w^{11}	—	—	—
—	—	—	w^{12}	w^{11}	w^{10}	w^9	w^8	w^9	w^{10}	w^{11}	w^{12}	—	—	—
—	—	—	w^{13}	w^{12}	w^{11}	w^{10}	w^9	w^{10}	w^{11}	w^{12}	w^{13}	—	—	—

Now note the two following facts:

a) The sum of all the numbers shown under the line is

$$S = (w^5 + w^6 + w^7 + \cdots) + 2(w^6 + w^7 + \cdots) + $$
$$+ 2(w^7 + w^8 + w^9 + \cdots) + \cdots =$$
$$= \frac{w^5}{1-w} + 2\frac{w^6}{1-w} + 2\frac{w^7}{1-w} + \cdots$$

and since $1 - w = w^2$, then $S = w^3 + 2w^4 + 2w^5 + \cdots =$
$$= (w^3 + w^4 + w^5 + \cdots) + (w^4 + w^5 + w^6 + \cdots) =$$
$$= \frac{w^3}{1-w} + \frac{w^4}{1-w} = w + w^2 = 1$$

$$\boxed{S = 1}$$

Thus, *the sum of the numbers corresponding to a finite number of squares is strictly less than 1.*

b) Now let's interpret our allowed moves with the coins with regard to these numbers as follows. We add the numbers corresponding to the squares of the board occupied by pennies before and after a move. For example:

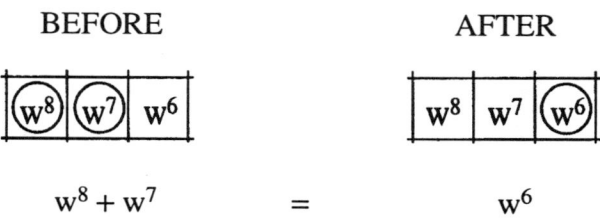

Another case is:

$$w^9 + w^8 = w^7 \quad > \quad w^{10}$$

This fact, as you can easily see, is general. In other words, *the sum of the occupied spaces is the same or smaller over the course of all our possible moves.*

Now, the sum of the spaces occupied at the beginning of the game (a finite number of occupied spaces) is, as we have seen, less than 1. If we could manage to get to the fifth row with our allowed moves, placing our 1 in the square we reach in the fifth row, it would turn out that the sum of the occupied spaces at the end of the game would be greater than or equal to 1, which is impossible.

Conway's idea is very ingenious. Why don't you try to use it to clear up some of the questions you still have. A little algebra will help.

1) Prove that the minimum number of pieces needed to reach row 3 is in fact 8, and that for row 4, it is 20. The second part of this problem is not easy.
2) Prove whether the only initial positions for reaching rows 3 and 4 are the ones shown and their symmetries.
3) Come up with other similar problems, like whether it is possible to reach these situations, and if so, how.

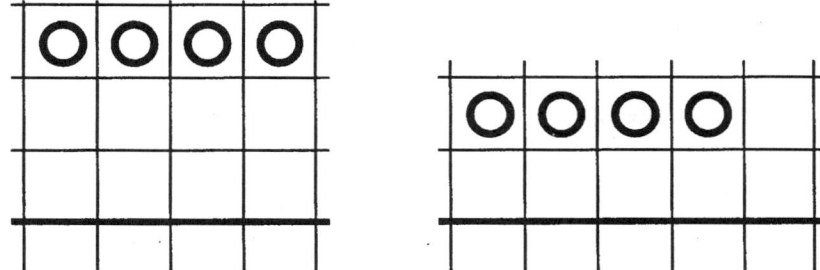

And I'd appreciate it if you'd write to me if you have any kind of promising idea about all this mess. Thanks.

Notes

The golden number, which appears in this chapter, has, since the time of the Ancients, been a mysterious number that appears in strangely diverse situations. The golden section of classical, neoclassical and modern monuments is the geometric proportion in which an inner point P of a segment AB divides the segment in such a way that $AB/AP = AP/PB$. Thus the value of AP/AB is 0.618, which is known as the golden number.

One of the curious situations in which the golden number appears is the following: Leonardo de Pisa, the most important mathematician of the Middle Ages (born around 1170), also called Fibonacci, in his most important work, *Liber Abaci*, where he introduced Arabic numbers in the West, proposed the following interesting problem: "A certain man put a pair of rabbits in a place that was completely fenced in.

How many pairs of rabbits will be bred starting with the first pair over the course of one year, assuming that each pair breeds a new pair each month, and the new pairs are productive from their second month?" The sequence of numbers corresponding to the number of pairs that there are in each of the successive months is 1, 1, 2, 3, 5, 8, 13, 21, 34, 55,..., in which each term is the sum of the two previous ones, is known as the **Fibonacci sequence**. Well, the ratio of each term in the Fibonacci sequence to the next one gets closer and closer to the golden number (the limit of this ratio is precisely the golden number).

Abridged Chess

This is another type of solitaire that's much more modern. It's also very odd. You can play it with a chess board and a set of dominoes. You'll need 31 dominoes, but since a set contains only 28 pieces, if you don't have two domino sets on hand you'll have to make three pieces of paper of the same size as the dominoes. Actually, we won't be using the numbering on the dominoes at all, so you can also use 31 pieces of paper, each measuring the size of two contiguous squares on the chess board.

Now take two coins and cover the squares of two opposite corners on the chess board. Now the idea is to cover all the other squares with the 31 dominoes, each covering two side-by-side squares on the chess board.

If you think I'm trying to trick you, try to prove that the game is impossible. I'll give you a hint. A chess board is quite large, with 64 squares. Why don't you try first with a smaller board, say one measuring 6 by 6, using 17 dominoes, or even a smaller one?

Now close your book and get to work. Try it for a while before you come back.

Yes, it's true. The game is impossible. If you began with the easiest case it probably didn't take you long to make this discovery.

After placing coins on two opposite corners, a four-square board looks like this:

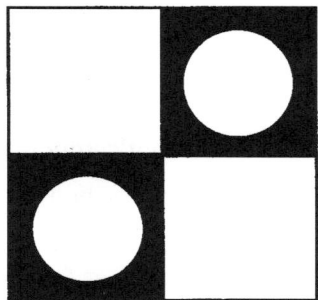

Clearly there is no way to cover the two remaining squares with a domino.

With a nine-square board, the game doesn't make much sense, since there are 7 squares left after we cover the two corners, and three dominoes would cover 6 squares while four would cover 8.

On the 16-square board

things are not so obvious, but the case of the four-square board may have clued you in. On starting, we have covered *two black squares* with coins. So there are 8 white squares and 6 black squares left to cover. But each domino must necessarily cover one black and one white square! So, if we were able to place the seven dominoes, we would have to cover seven white squares and seven black ones. *This solitaire game is impossible.*

The case of 64 or 100 squares is exactly the same. As you can see, sometimes starting with the simplest examples can clarify a lot.

To make things more complicated, we're going to prove the same thing using the Klein group, following the idea of the solitaire game invented by the man from the Bastille. We write the following elements of the Klein group on the chess board:

a	b	a	b	a	b	a	b
b	a	b	a	b	a	b	a
a	b	a	b	a	b	a	b
b	a	b	a	b	a	b	a
a	b	a	b	a	b	a	b
b	a	b	a	b	a	b	a
a	b	a	b	a	b	a	b
b	a	b	a	b	a	b	a

The product of all the elements is $a^{32}b^{32} = 1$. If we remove two opposite corners (or two squares of the same color, like the two squares marked b), the product is also 1, since it is $a^{32}b^{30} = 1$. If we interpret the placement of a domino as a division by the elements of the group that the domino covers, to place a domino is to divide by ab. Thus, to place 31 dominoes is to divide $a^{32}b^{30}$ by $a^{31}b^{31}$, which is to say that the result is ab. But if we had covered all the squares exactly, the product would have to be 1. Thus the game is impossible. In other words, *if the initial coins are placed in two squares of the same color, the solitaire game is impossible.*

But what if we initially use the coins to cover two squares of different colors? Then none of our reasoning demonstrates this version of solitaire to be impossible. With the Klein group reasoning, we are dividing by ab, and thus we are left with $a^{31}b^{31} = ab$. To place 31 dominoes is to divide by $a^{31}b^{31}$, leaving 1, which is what we should end up with to have the board completely covered.

Now will the solitaire game we've proposed be possible or not? Why don't you make a few attempts before reading on, experimenting with a few specific cases? Do it on small boards. If it is possible, can you offer a solution for solving this solitaire game when it can be done?

The solitaire game we've proposed here is possible, and here are two solutions. The first one is mine and the second belongs to R. Gomory, a mathematician at IBM.

Suppose one of the covered squares, c_1 or c_2, is not on one of the edges of the board (the case where both are on the edges is easy, following the same lines, and I'll leave it for you to solve). Let that square be c_1. Choose a region of the board determined by c_1 and one edge of the board such that c_2 is not located within the region, as shown below:

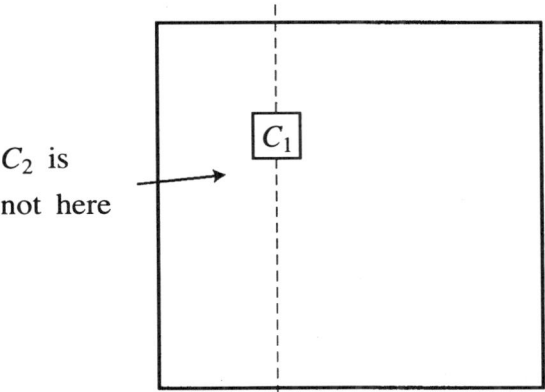

Now choose the winding path I've indicated along the squares of the board to a corner.

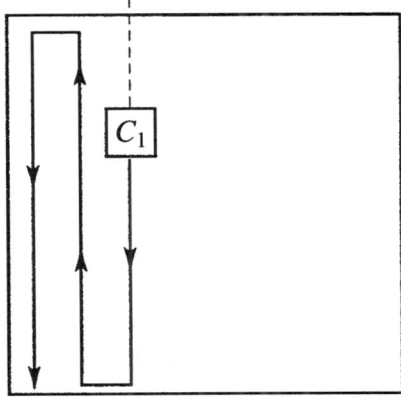

This path goes through an even or odd number of squares. If the number is odd, then the following opposite path

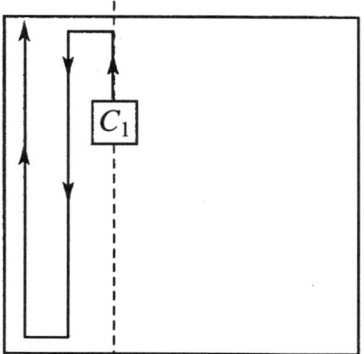

goes through an even number of squares. We choose the path with an even number of squares and fill it with dominoes in order, beginning with the square adjacent to c_1; since there is an even number of squares, it can be done. Suppose that the even path is the second one shown above. Now we choose the following winding path:

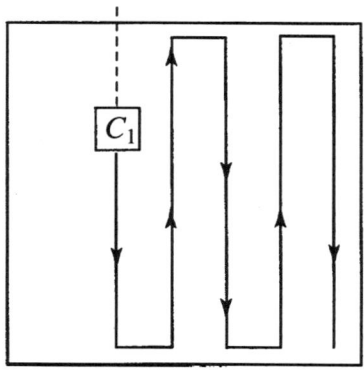

We begin to cover it with dominoes beginning with c_1 in the direction shown. Square C_2 is on the path, but because of the difference in color, we finish covering the squares evenly up to the square next to c_2. Then we skip c_2 and cover the rest of the board with no hitches. The reason underlying this solution is clear. Note that there is an even number of squares from c_1 to c_2, no matter which path you take; this is also true from c_2 to the last corner square.

Gomory's solution is as follows. We imagine the board covered with the two Gomory forks I've shown below ("wiggly" thick lines):

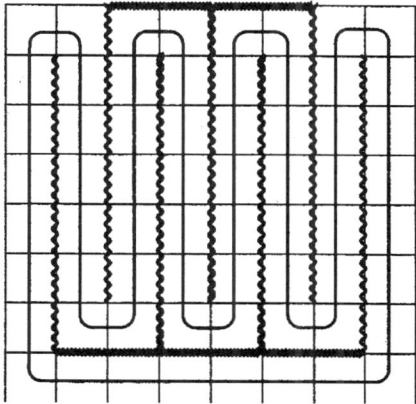

What the two forks do is to mark the closed path indicated on the figure that visits all the squares just once. If you cover one white square and one black one in any place, it is possible to fill the path between them in both possible directions, since there is an even number of squares in either one. Pretty clever, huh?

What other problems of this type can you think of? For instance, cover up one or three squares on a 9 by 9 board. Can the remainder be filled in with dominoes? How?

Notes

The game of chess suggests a multitude of puzzles and games of a wide variety. One of them, interesting and very old, which allows a complete mathematical treatment is the eight-queens game. Can you place eight queens on the board without any of them being in a position to take any of the others? More generally, on an n by n board, what is the maximum number of queens that can be placed on the board without having one threaten any of the others? How many different possibilities are there for placing them with this condition? An interesting game for two players, A and B, based on this problem, is as follows: A places a queen somewhere on the board, then B does the same thing in such a way that his queen is not threatened by the first one, then A has to place another one such that this queen is not threatened by any of the previous ones. The loser is the one who is unable to place a queen without putting it in danger of being captured.

Another game that also goes way back is the knight's stroll. Can you mark out a path for a knight to visit all the squares of the board without having it visit the same square twice? Can you do it starting at just any square? What about any board measuring n by n or n by m?

The Secret of the Oval Room

The great Oval Room was filled with spies, counterspies and counter-counterspies. And yet the Prime Minister was in absolute need of informing His Majesty immediately of the great secret he had just learned. So, cool as a cucumber, when he approached the King he said quite audibly, "Your Majesty, it would seem that the rebel outbreaks require our attention." All the spies moved towards the walls of the room to take their coded message keys out of the linings of their capes.

Naturally, following them with great stealth were the counterspies, who were, in turn, followed by the counter-counterspies. The King, walking calmly but determinedly, started off towards one side of the Oval Room. The Minister, for his part, with the same determined but calm gait, walked off towards the other side of the Oval Room. The spies watched them from the corner of their eyes as they checked out the meaning of "seem," "rebel," "outbreaks" and "require" in their booklets. The attention of the counterspies was focused on the spies, and the counter-counterspies weren't letting the counterspies out of their sight for one second. The King stopped for a moment and the Minister, respectfully, also came to a halt. They were more than 70 feet apart when one of the sharper spies observed and wrote in his booklet: "This Minister is either talking to himself or praying." But no one could hear a thing. Only the King was able to hear the words of the Minister: "Your Majesty, with all due respect, your fly is wide open."

The mystery of the Oval Room consists fundamentally of the fact that in an ellipse such as this one:

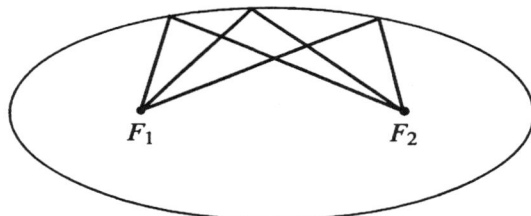

there exist two points, F_1 and F_2, the foci, such that if the walls of the ellipse were made of rubber like those of a pool table and a ball were thrown from F_1 in any direction, on bouncing off the wall it would end up passing through F_2. When sound bounces, it behaves like the ball. That's why when you speak very, very softly at F_1, your voice reaches F_2 with sufficient intensity to be understood, since it reaches F_2 from all the directions that depart from F_1. At any other given point, only the sound aimed at that point actually reaches it and it is not perceptible.

Sometimes there are subway stations that have a section with a ceiling that is more or less elliptical. Try this experiment. Have a friend stand on the opposite platform in such a way that when you whisper a secret message, she can hear you. Of course, make sure you don't try this just as one of those infernal monsters is charging through. The experiment doesn't usually work at those moments!

So who was it that said that ellipses have this wonderful property? You already know it — we saw it earlier. Remember "The Mathematics of a Sandwich"? But allow me to be a bit tedious and repeat myself.

With the center at F_1, draw the circumference of radius $2a$, the length of the biggest axis of the ellipse. Take a point M of the ellipse and join it to F_1 and F_2. Since $MF_1 + MF_2 = 2a$, we see that if we extend MF_1 to N on the circumference, $2a = MF_1 + MN$, and thus $MN = MF_2$. If you draw t, the mediatrix of NF_2, you find that t has a common point M with the ellipse. Could it have more? No, because $SF_1 + SN > NF_1 = 2a$ (one side of a triangle is less than the sum of the other two). So t is the tangent to the ellipse at M. Now, when a ball bounces off a curve, it bounces as if the curve were substituted by its tangent.

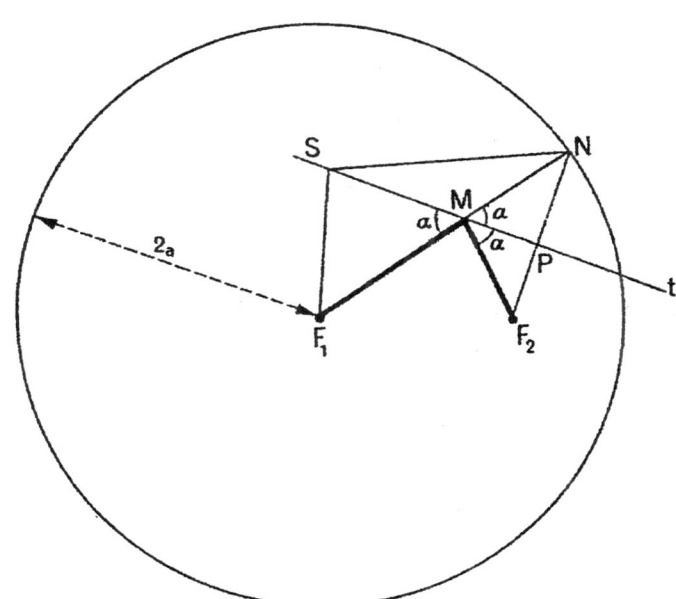

A ball thrown from F_1 to M would bounce off at t forming equal incoming and outgoing angles with t. But since the angle F_1MS is equal to the angle NMP (opposites at the vertex) and NMP is equal to PMF_2 (symmetry with respect to t), it is clear that the ball traveling from F_1 to M will bounce towards F_2.

Did you know that the most powerful telescopes have an enormous parabolic mirror? Why? The rays of light that reach us from a star are virtually parallel to each other and are gathered by the parabolic mirror with its axis placed in the direction of the rays, and when reflected they end up at the focus of the parabola. Interesting, huh? Why is this? The light that comes from a star is very faint. If all the light that comes through the big opening of a parabolic telescope is gathered, there is much more. How is that? It's similar to the ellipse problem. A parabola is the set of points on the plane equidistant from a point, the focus, and from a line, the directrix. The idea is to see how ray r is reflected parallel to the axis when it meets at a point M of the parabola.

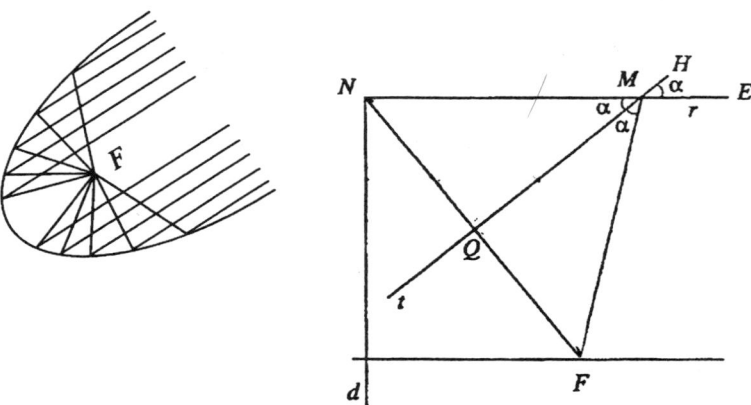

To do this we have to find out what the tangent at M to the parabola is like. Since $MN = MF$, the mediatrix of NF goes through M. Does it have any other point in common with the parabola? No, because $SF = SN > SP$ (hypotenuse greater than the leg). Thus t is the tangent at M. Now, angle EMH = angle NMQ = angle QMF, and so ray r comes to be reflected towards F.

Lewis Carroll, the author of *Alice in Wonderland*, was very fond of puzzles and mathematical games, as you can guess from the writing style of *Alice in Wonderland*. They say that when the book was published, Queen Victoria of England liked it so much that she asked that a copy of whatever the author published be sent to her. The next publication she received was... a treatise on geometry! Lewis Carroll, the pseudonym of Charles L. Dodgson, was a professor of Mathematics. Among other ingenious devices, Carroll put together an elliptical billiard table. As you can easily imagine, it would have to be interesting to play on an elliptical billiard table. Strange things occur. A ball that starts at focus F_1 without any spin goes through the other focus, F_2, then through F_1, then through F_2, etc. Furthermore, the ball's direction approximates more and more that of the major axis. On the other hand, if a ball goes through one point of the segment that joins the foci without going through them, it will always (assuming it doesn't stop) pass between the foci, and the envelop of the resulting trajectory is a hyperbola with the same foci. But if the ball at some point cuts the major axis between one focus and the point closest to it on the ellipse, it will always do so between the foci and the ellipse. Why don't you try and prove it? It's fairly easy with what you already know.

Here's another interesting problem to think about. If you solve it, by all means go ahead and jump for joy, because although the problem looks simple, it's been floating around for many years with no one able to come up with the solution.

Since we're working with strange billiard tables, let's make one more, this one polygonal. The question is: Does there exist a polygonal billiard table such that there are two points on it, *M* and *N*, located in such a way that a ball placed at *M*, no matter how it is thrown, can never reach point *N*? Naturally, it is assumed that the ball never stops.

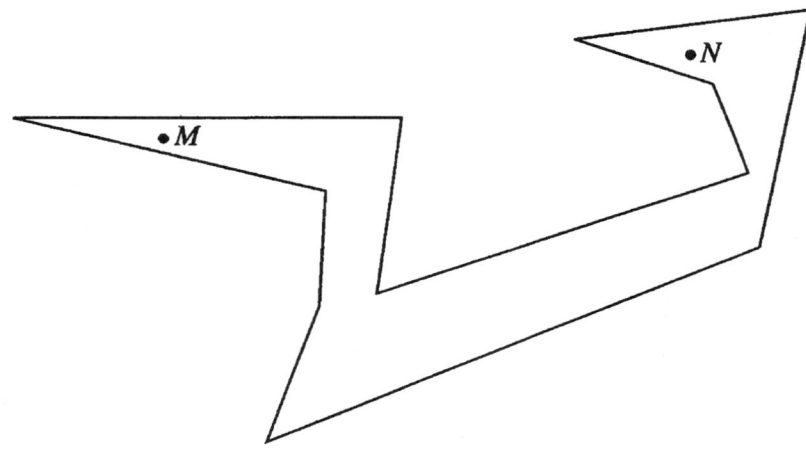

A curved billiard table with this property can easily be built as follows, based on the properties of ellipses that we know:

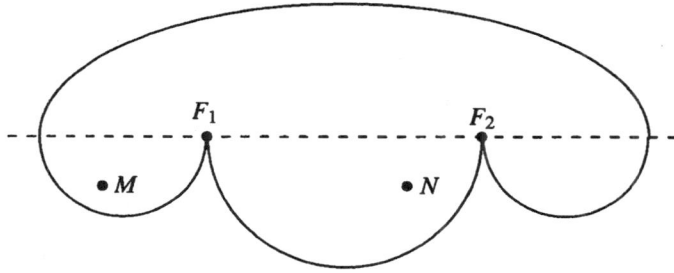

The upper part is half of an ellipse with foci F_1 and F_2. The other parts of the edge are semicircular. If M and N are placed as shown, no matter how much the ball rebounds, if it leaves M it will never meet any point of the semicircle whose diameter is F_1F_2. If only straight lines are allowed (a polygonal billiard table), it is still not known whether there exists a table with a similar property. Why don't you give it some thought?

Notes

One of the great mathematical geniuses of all times, Blaise Pascal (1623–1662), began his career at an early age with a famous work on conical sections. Pascal was born in Clermont, France, where his father was a magistrat. His mother died in 1626, and from that point on, the father took personal charge of the education of his two children, Blaise and Jacqueline, born in 1625. Jacqueline showed talent early on in the literary field, while Blaise did so in mathematics. At the age of 16, he wrote a treatise on conical sections that aroused the admiration of Descartes, who was never fully convinced that it could have been the work of an adolescent. Pascal's scientific genius is astounding. Among his many outstanding results are his ideas on complete induction, the invention of the principles of the calculus of probabilities, the discovery and utilization of the principles of calculus and the construction of what was probably the first calculator. Around 1646 he began to get interested in the principles of hydrostatics, atmospheric pressure and vacuums, in many areas going beyond Torricelli, one of his contemporaries.

But Pascal came to be even more famous as a philosopher of religion and a writer, putting him in a unique place in the history of thought. Before he turned thirty, Pascal had secured incomparable renown as a mathematician and physicist. Beginning in 1654, after a deep religious experience he called his "night of fire," he decided to devote himself fully to a life of prayer and religious thought.

His most important works in this aspect are The Provincial Letters, of great influence in their time, and especially all the fragments found following his death of the great work he was planning in the form of an apologia of the Christian religion.

Bibliography

As a farewell, here are the titles of a few books you may like, written with the same overall aim in mind as this book.

Abbot, E. A., *Flatland: A Romance of Many Dimensions*, Princeton University Press, Princeton, NJ, 1989.

Gardner, M., *Aha! Gotcha: Paradoxes to Puzzle and Delight*, WH Freeman & Co., New York, 1982.

Kline, M. (ed.), *Mathematics in Western Culture*, Oxford University Press, 1965.

Rademacher, H. and Toeplitz, O. *The Enjoyment of Mathematics*, Dover Publications, 1990.

Smullyan, R., *What Is the Name of This Book?: The Riddle of Dracula and Other Logic Puzzles*, Prentice-Hall, Inc., 1978.